SCALE MODELER'S HOW-TO GUIDE

CLASSIC COMBAT AIRCRAFT Vol. 2

MODELING WWII WARBIRDS

Compiled by Jeff Wilson
Introduction by Paul Boyer

KALMBACH
BOOKS

Printed in the United States of America

11 10 09 08 07   1 2 3 4 5

Visit our Web site at kalmbachbooks.com. Secure online ordering available.

Cover photo
Alfonso Martinez Berlana's 1/48 scale model of WWII ace Francis "Gabby" Gabreski's P-47D was photographed by Aurelio Gimeno Ruiz.

The material in this book has appeared previously as articles in *FineScale Modeler* magazine.

Publisher's Cataloging-In-Publication Data
(Prepared by The Donohue Group, Inc.)

Classic combat aircraft. Vol. 2, Modeling WWII warbirds / compiled by Jeff Wilson ; introduction by Paul Boyer.

    p. : ill. ; cm. -- (Scale modeler's how-to guide)

  Volume 1 published as: Modeling classic combat aircraft / compiled by Mark Thompson. 2003.
  "The material in this book has previously appeared as articles in FineScale modeler magazine."
  ISBN: 978-0-89024-696-2

1. Airplanes, Military--Models--Design and construction. I. Wilson, Jeff, 1964- II. Title. III. Title: Modeling WWII warbirds IV. Title: FineScale modeler.

TL770.A2 C53 2007
623.74/6/0288

Introduction

Projects

Contents

M ore than 60 years have passed since the end of World War II, and still that conflict, and the machines used to fight it, are the predominant interest areas among scale modelers.

World War II remains a primary magnet drawing modelers to the hobby – even young ones like Guntam Rao, 16, who expressed his interest in the war's beginnings with this 1/72 scale Tamiya Spitfire Mk.1 in service during the Battle of Britain. FSM staff photo

Perhaps it is the lore, perhaps the legend, perhaps the livery, but most likely it is the sum of all these along with the vast variety of aerial weapons that makes WWII the most popular modeling category. The need for speed, better weapons, and greater manufacturing capacity drove the industry forward to create the classic Spitfire and Messerschmitt, Mustang and Zero, Macchi 202 and Stormovik. Think of it: World War II started with most of its participants flying 200-mph biplanes in some frontline units, and the war ended with several air arms flying 500-mph jet fighters. From Gladiators to Meteors, from He 51s to Me 262s, from Willows to Kikkas, the design and engineering progress of aviation in the early 1940s is unmatched.

There was a time in the 1980s when interest in WWII aircraft models waned. I remember a kit manufacturer's rep telling me "WWII is dead. The only thing that sells is jets and helicopters." But in the late 1980s, WWII surged back in front with new kits coming from Asia and Eastern Europe. The lifting of the Iron Curtain and new trade agreements with China opened new markets and brought in new manufacturers and innovations.

Since then, manufacturers from around the world have supplied modelers with hundreds of different kits, to the point where you wouldn't have to convert a Messerschmitt Bf 109G-10 to a G-6 – you would just go out and buy a G-6 kit. Want to model McCampbell's "Minsi II" instead of the "Minsi III" Hellcat? No problem, just buy a different decal sheet. Detail freaks can feed on a variety of aftermarket detail sets. If you're not hungry for detail, you can dine on a seemingly infinite variety of color and markings.

Within these pages, you'll find some of the best WWII aircraft models ever to appear in *FineScale Modeler* magazine. Modelers from around the globe have found the universal appeal of WWII subjects, and they're willing to show fellow modelers their work. You're sure to find tips, tools, and techniques to help you model your favorite WWII aircraft. And then the next one. And then another. And … .

– *Paul Boyer*

Paul Boyer retired as Senior Editor of FineScale Modeler *in 2006 after a career of nearly 25 years with the magazine.*

King of the STRAFERS

James Goodson's first P-51B — Converting and superdetailing Hasegawa's 1/32 scale bubbletop Mustang BY PAUL BUDZIK

The business end of Paul's Mustang shows off James Goodson's personal emblem and impressive scoreboard. Real or model? Master modeler Paul Budzik performed miracles on this "razorback" P-51B Mustang converted from Hasegawa's 1/32 scale bubbletop "D." In addition to the conversion work, Paul added loads of scratchbuilt details, and dropped the flaps and elevators, too.

When I started gathering reference material for my next project, I looked for an interesting camouflaged 4th Fighter Group P-51B Mustang. I've always admired James Goodson, and after looking over my references, I figured that Goodson must have flown a camouflaged P-51B between his time in a razorback P-47D and his well-known natural-metal P-51B.

Goodson's son Jamie helped me with more references and a valuable lead: Goodson's crew chief, Bob Gilbert, who might have a photograph of an earlier B model in camouflage. Sure enough, several weeks later Gilbert sent a photograph of Goodson's first P-51B, serial no. 43-7059. Fantastic!

Starting point. There are no perfect kits, but I determined that trying to improve

Revell's old 1/32 scale P-51B would be more trouble than converting Hasegawa's P-51D bubbletop. The conversion involved the canopy, razorback, and wings, but I also improved detail overall with plenty of scratchbuilt parts.

I can't give you a blow-by-blow account of all the improvements to this model, but you may find some of the techniques shown here helpful on your next project.

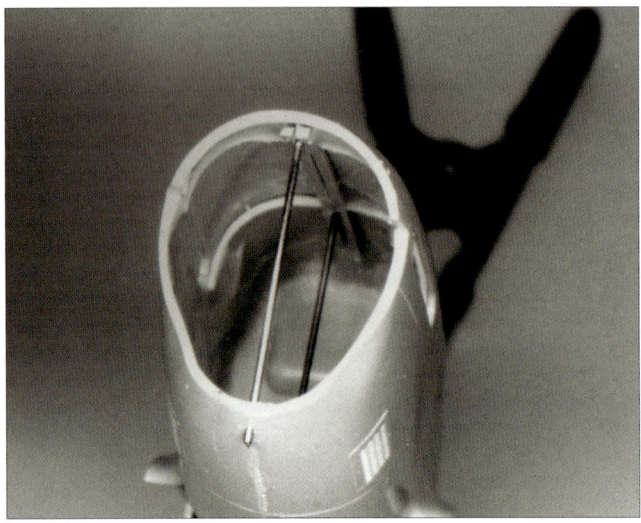

The Hasegawa kit fuselage halves were warped because of the separate cowl panels and the opening for the wing. Paul inserted .032" wire on the nose center line to serve as a sight to align the fuselage with the vertical fin.

With the aid of a template made from a 1/32 scale drawing, Paul fashioned a pattern over which he vacuum-formed several clear acrylic canopies.

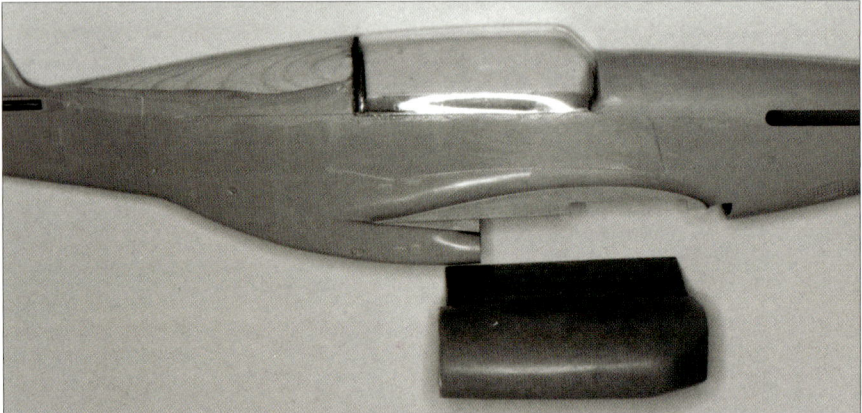

After the canopy was cemented in place, the rest of the spine was carved from wood, filled, and faired in. Paul uses a homemade filler of super glue and dental resin powder. The clear canopy was then removed and work started on the cockpit.

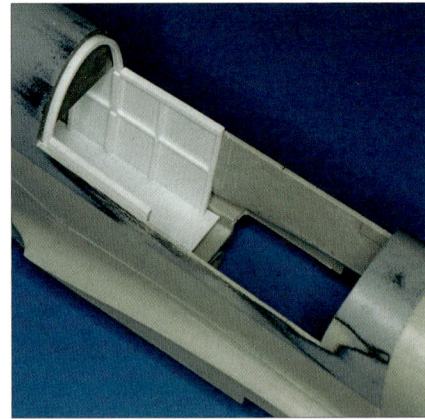

The first step in detailing the interior was adding new cockpit formers and stringers made of strip styrene.

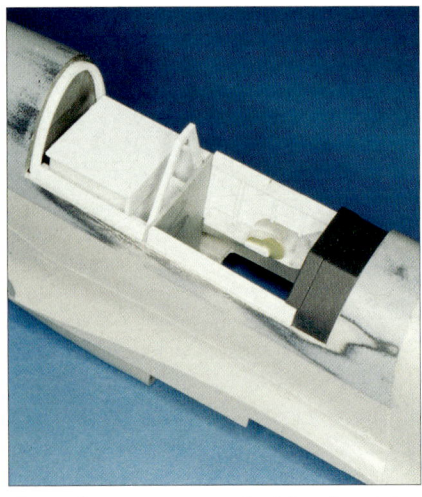

The radio compartment bulkhead and the radio shelf were made from sheet styrene. Later, Paul milled a radio from a chunk of acrylic.

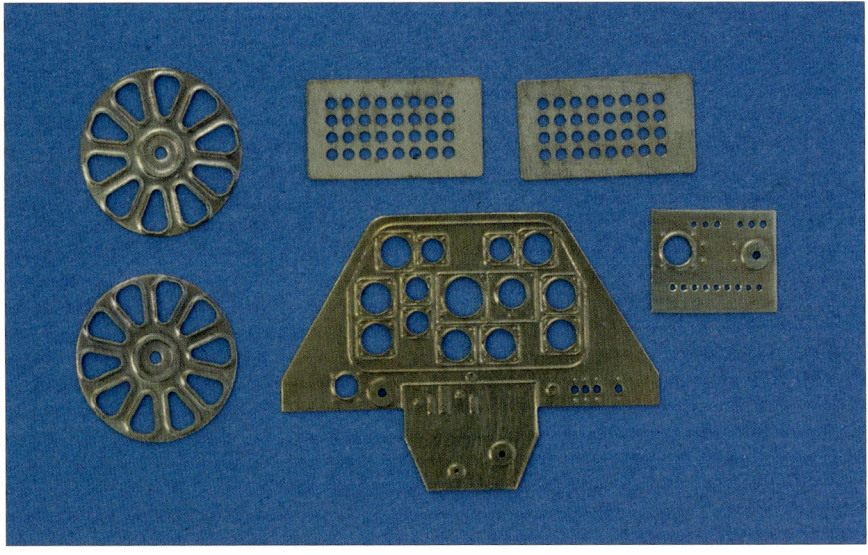

Paul produced his own photoetched brass parts. Here are the instrument panel, radio control panel, breather grilles, and main wheel faces.

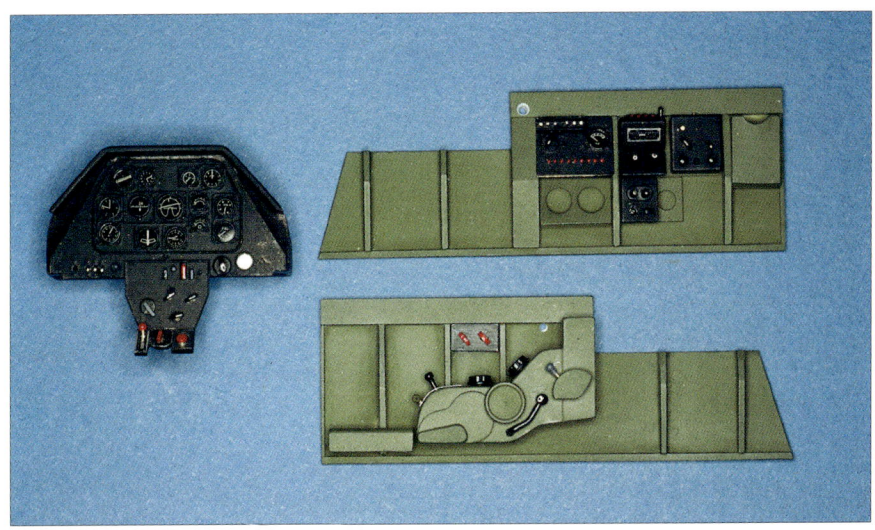

The Mustang's finished multilayer instrument panel is at left. The right (top) and left cockpit interior panels were made from styrene and milled acrylic shapes.

The finished panels were inserted into the cockpit through the opening in the bottom of the fuselage. Yet to come are the cockpit floor, seat, and control stick.

Next, Paul reinstalled the clear canopy, faired it in, and cut away the open portions.

The scratchbuilt seat was made from sheet styrene with surgical-tape seat belts and Waldron Products' photoetched hardware.

The cockpit floor and rudder pedals were made from sheet styrene, while the blocks that hold the pedals and the control-stick boot were carved from acrylic.

JAMES GOODSON, KING OF THE STRAFERS

Combining his skill at low-level flying with a penchant for destroying enemy aircraft, James Goodson earned the title "King of the Strafers." He was tough in dogfighting as well, ending the war with 15 victories in the air and 15 destroyed strafing.

Although an American, "Goody" first enlisted with the Royal Canadian Air Force and was posted to an RAF squadron flying Hurricanes and Spitfires from England. He was transferred to 133 Squadron, one of the Eagle squadrons of American volunteers. The American fliers were absorbed into the USAAF as the 336th Fighter Squadron, 4th Fighter Group, in September 1942.

Goody proved to be an excellent leader and rose to the rank of major, commanding the 336th Squadron. At war's end, his decorations included a Presidential Unit Citation, the Distinguished Service Cross, Silver Star, Distinguished Flying Cross with eight clusters, the Air Medal with 20 clusters, and the Purple Heart.

After the war, Goodson received his MBA from Harvard University. Fluent in four languages, he rose to head both Goodyear and Hoover in Europe. Later he accepted a position as vice president and group executive at ITT.

Above: Goodson (right) points out his latest victory symbol to Ralph "Kid" Hofer. This aircraft was Goodson's later P-51B that featured a natural-metal finish and the "Malcolm" hood canopy. Photo courtesy of Roger Freeman and Jamie Goodson

Left: The only known photo of Goodson's camouflaged P-51B. Photo courtesy Bob Gilbert

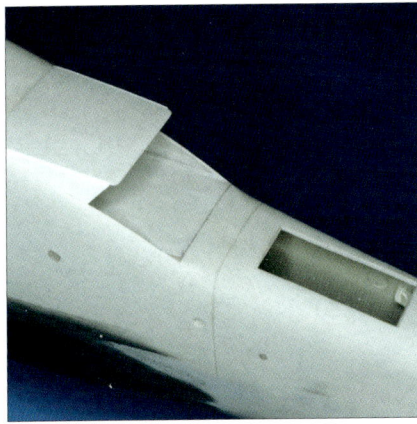

Paul replaced the radiator outlet doors with sheet styrene.

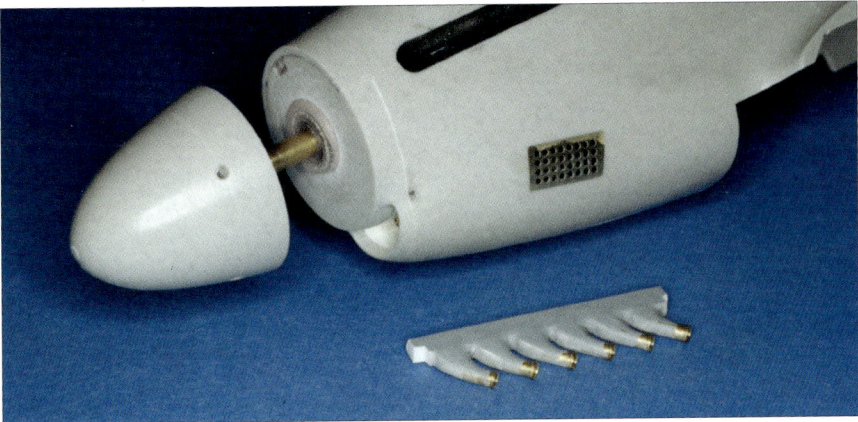

Modifications to the nose included lining the shallow chin scoop with sheet styrene and replacing the shrouded exhaust stacks with new ones machined from styrene and tipped with turned brass. Paul cut away the kit spinner mount and inserted a new one made of acrylic. A hole drilled in its center holds a brass tube attached to the spinner. Note the photoetched breather grille.

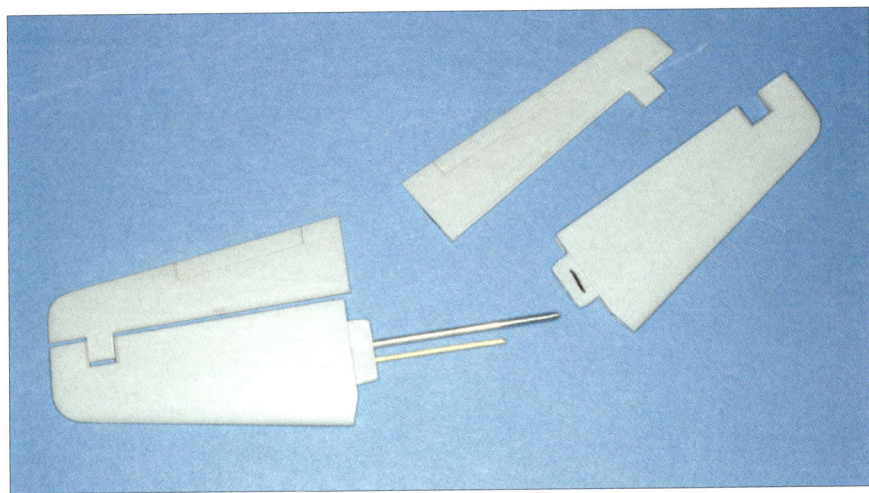

Paul separated the elevators from the horizontal stabilizers, refined the hinge lines, and reinserted them with a slight nose-down deflection. To align the stabilizers, he inserted wire pins into the mounts.

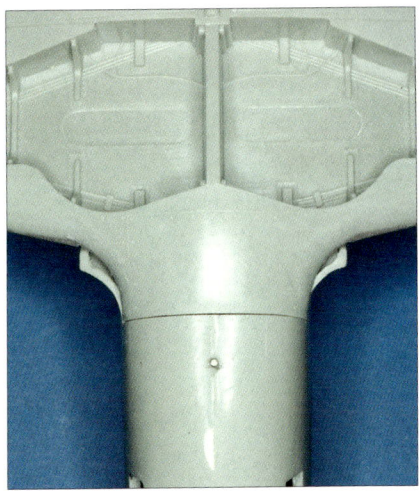

Paul had to reduce the leading edge fillets to represent the B model's wing. You can see the stubs of the D model molded on the kit fuselage. These were also cut back and smoothed out.

Dropping the flaps on the model meant rebuilding the exposed surfaces of the wings, flaps, and fuselage.

A silicone rubber mold was made to cast clear polyester-resin copies of a scratch-built acrylic main gear tire.

After milling out the center of the tire castings, Paul inserted scratchbuilt wheels made of brass castings and photoetched spokes.

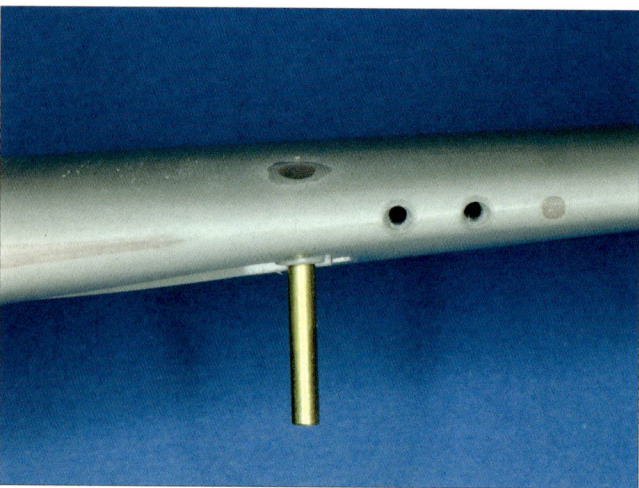

Each scratchbuilt main gear strut fits into a brass tube embedded in the wing. Note that the tube had projected through the top of the wing and has been filled and sanded smooth. You can see the refined machine gun ports; the B model had only four guns. Paul filled the outboard opening on each wing along with its associated shell ejection port underneath the wing.

Here's the finished strut and wheel; each strut was machined from brass rod. Note the drop tank plumbing made from solder and wire.

A peek into the finished cockpit makes you feel like you're looking at the real thing. Most of the markings were painted on, but Goodson's diving-eagle emblem came from an old IPMS/USA decal sheet – it was meant for a 1/48 scale P-47D razorback, but it fits the 1/32 scale Mustang, too.

AIR SUPERIORITY IN EUROPE – THE FAMOUS 4TH FIGHTER GROUP

The 4th Fighter Group was formed from the three Eagle Squadrons of the Royal Air Force. As such, it automatically became the oldest unit in the new Eighth Air Force and the only group to be activated in a combat zone. The 4th had the highest score of combined air and ground victories of any Eighth Air Force unit, was the first unit over Paris, the first to penetrate German air space, and the first over Berlin.

As P-47s arrived in England, the 4th traded its beloved Spitfires. Then, in February 1944, the 4th was one of the first groups to receive P-51s.

Under the leadership of Col. Don Blakeslee, the 4th enjoyed its greatest successes. Such pilots as Beeson, Gentile, Glover, Godfrey, Goodson, Hively, Hofer, McKennon, Millikan, Norley, and many more became high-scoring aces. Between March 5 and April 24, 1944, the 4th Fighter Group destroyed 189 German planes in the air and 134 on the ground.

Paul primed his model with lacquer, adding a second coat on certain panels to produce a slightly raised effect. He used Floquil paints for the color coats, starting with white for the ID bands, red for the spinner, and home-brew mixes for Neutral Gray and Olive Drab. Note the freshly overpainted ID band on the fin.

REFERENCES

Aircraft Profile – North American P-51B & C Mustang Richard Atkins, Profile Books Ltd., Berkshire, England, 1982

P-51 Mustang Classic Aircraft No. 3 Roy Cross and Gerald Scarborough, Patrick Stephens, London, 1973

P-51 Mustang in Action Larry Davis, Squadron/Signal Publications, Carrollton, Texas, 1981

P-51 Mustang in Color Larry Davis, Squadron/Signal Publications, Carrollton, Texas, 1982

Fighter Command Jeffrey L. Ethell and Robert T. Sand, Motorbooks International, Osceola, Wisconsin, 1991

Flight Manual Mustang P-51B & P-51C North American Aviation Inc., Inglenook, California

Camouflage & Markings – North American P-51 & F-6 Mustang Roger A. Freeman, Ducimius Books Limited, London

The Mighty Eighth Roger A. Freeman, Doubleday and Company, Garden City, New York, 1970

Mustang at War Roger A. Freeman, Doubleday and Company, Garden City, New York, 1974

Escort to Berlin Gary L. Fry and Jeffrey L. Ethell, Arco Publishing, New York, 1980

Tumult in the Clouds James A. Goodson, Harrup Press, Canterbury, England, 1983

P-51 Mustang William N. Grant, Gallery Books, New York, 1980

US Army Air Force Fighter Part 2 William Green and Gordon Swanborough, Arco Publishing, New York, 1978

P-51 Mustang Robert Grinsell, Crown Publishers, New York, 1980

1000 Destroyed Grover C. Hall, Aero Publishers, Inc., Fallbrook, California, 1978

P-51 Mustang Aces William N. Hess and Thomas Ivie, Motorbooks International, Osceola, Wisconsin, 1992

The P-51 Mustang Len Morgan, Arco Publishing, New York, 1963

Aces of the Eighth Gene B. Stafford and William N. Hess, Squadron/Signal Publications, Carrollton, Texas, 1973

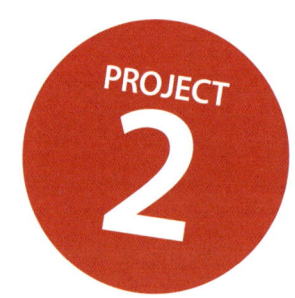

Gabby's LAST JUG

PROJECT 2

Detailing and painting Hasegawa's 1/48 scale P-47D bubbletop BY ALFONSO MARTINEZ BERLANA

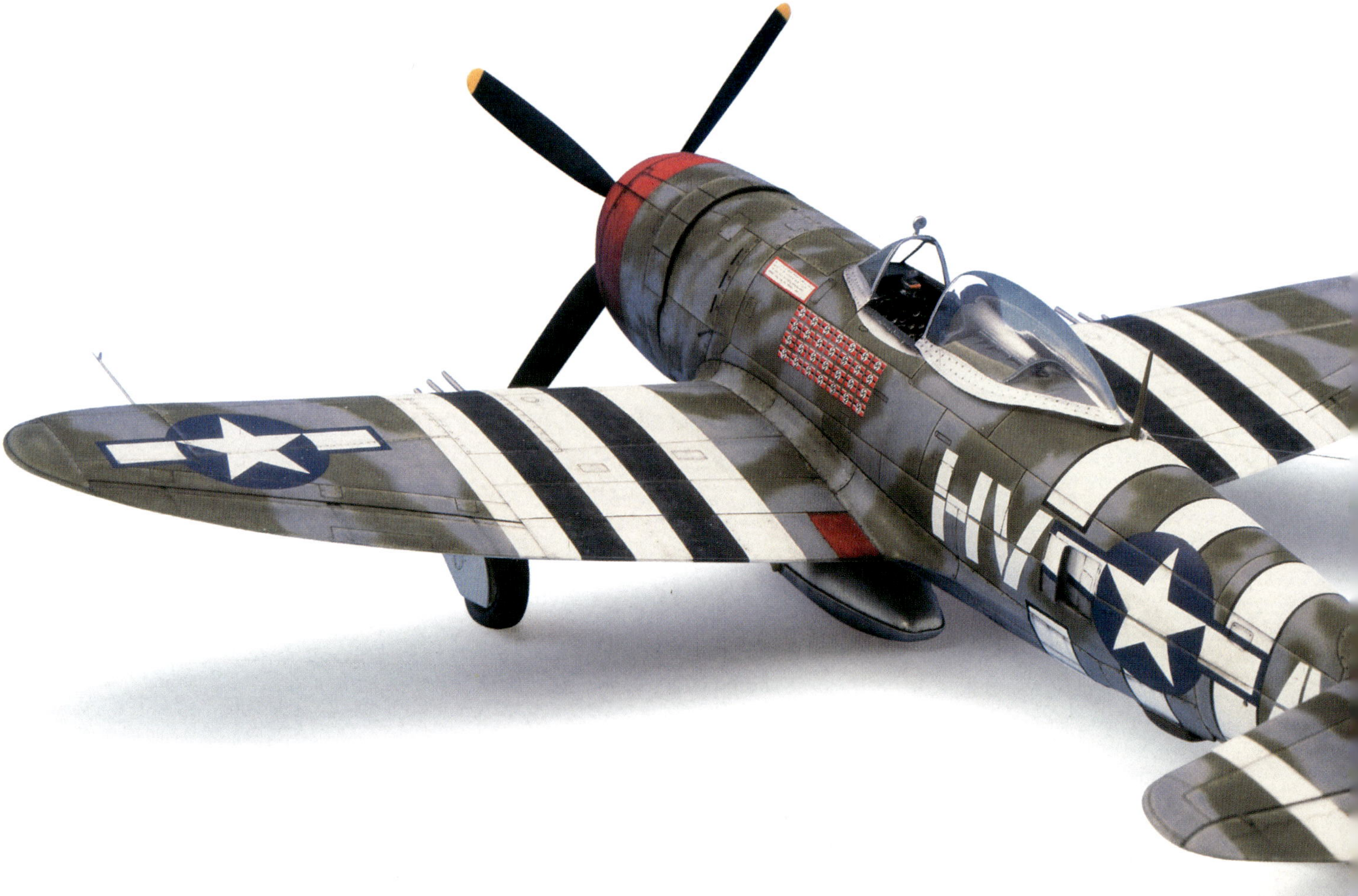

The unusual camouflage of Francis Gabreski's P-47D bubbletop Thunderbolt never looked better than on Alfonso Martinez Berlana's Hasegawa model. Aurelio Gimeno Ruiz photo

The passing of Francis "Gabby" Gabreski in January 2002 brought Hasegawa's bubbletop Thunderbolt to the top of my "to do" list. I just had to model Gabreski's famous field-camouflaged P-47D.

I chose Hasegawa's kit because it was well-detailed, fit nearly perfectly, and was easy to build. I found only one error that had to be corrected: The supports for the belly tank were too long. If they were placed as intended, the tank would have touched the ground. I cut them down so the tank would sit properly.

As good as Hasegawa's kit was, I just couldn't leave well enough alone. I added a True Details resin cockpit (No. 48483) and wheels (48009) and installed more detail on the engine and landing gear.

To this day, it is uncertain whether the bottom of Gabreski's P-47 was camouflaged, but the fuselage and upper surfaces appear to have been painted in RAF medium sea gray and dark green. I decided to go with the unpainted natural-metal bottom.

Let me show you how I detailed and painted Gabby's Jug.

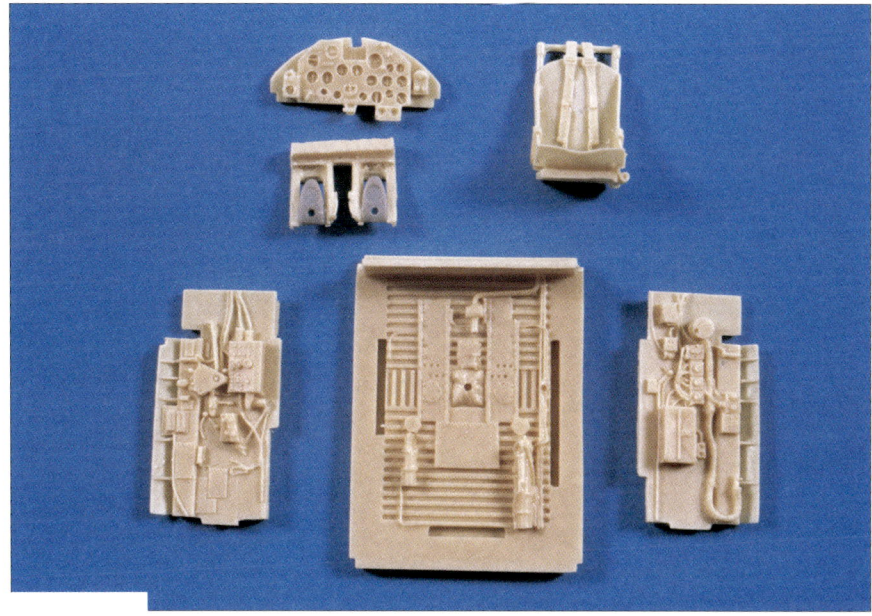

Here is the True Details resin Thunderbolt cockpit set unpainted (top) and how it looks after painting. I hand-brushed the interior with Vallejo acrylics. Gabreski's aircraft was a P-47D-25RE built at Farmingdale, N.Y. The cockpits of this production batch were painted in dull dark green, rather than the typical "interior green." Project photos by Alfonso Martinez Berlana

Right: I used thin copper wire and fine red plastic insulation stripped from thin-gauge electrical wire to form hydraulic brake lines on the landing gear struts. Short pieces of the tubing also form the connectors to the engine's ignition harness.

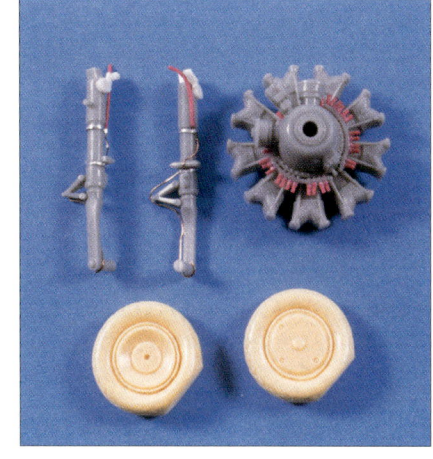

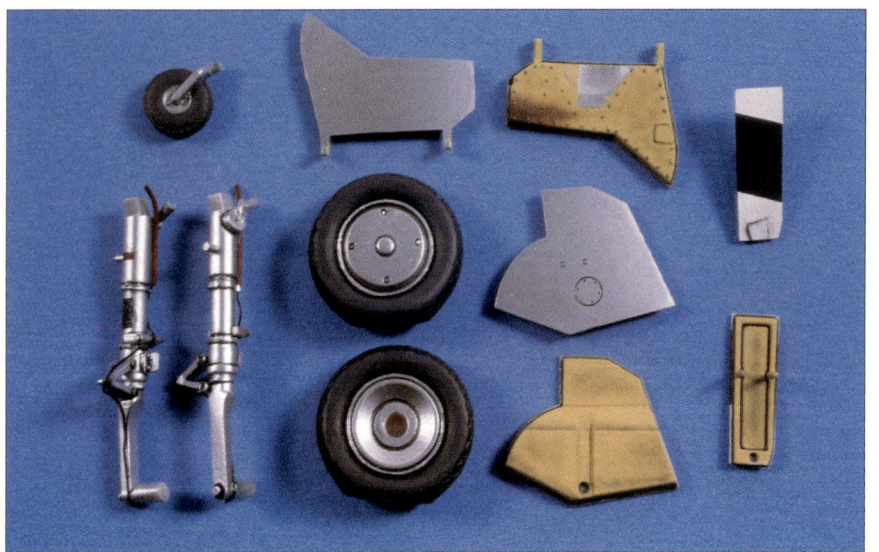

Here's the finished landing gear ready to install. The interior of the Thunderbolt's wheel wells was painted in yellow zinc chromate.

I used a No. 76 bit to drill depressions inside the clear wingtip navigation-light lenses. The depressions represent the place where the colored bulbs under clear lenses would be, so the right hole was painted with clear green, the left with clear red.

Going back to the engine, I wired the ignition system with fine copper wire and painted it while it was attached to the kit firewall.

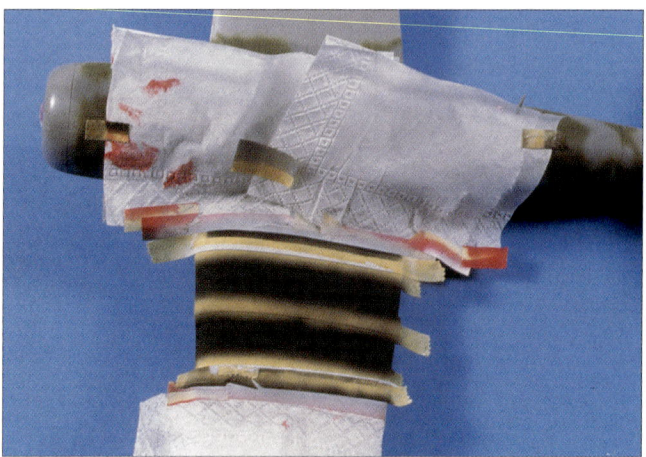

I allowed the camouflage to dry for a couple of days, then masked and painted the wing and lower fuselage invasion-stripe areas white. After the white was dry, I masked and airbrushed the black stripes. For more painting techniques, see Larry Schramm's article on page 38.

Before painting the camouflage, I polished, attached, and masked the windscreen on the fuselage, then sealed the cockpit and wheel wells with tape and liquid masking agent. A wipe-down with alcohol removed any oil from fingerprints.

I painted the bottom surfaces and the belly tank first with aluminum. I painted the upper surface camouflage with Tamiya acrylics, mixed roughly to the RAF medium sea gray and dark green that were probably used on the real aircraft. I lightened both colors with white for "scale effect."

I couldn't find pictures of the right side of the plane, so made up a likely pattern. To prevent paint buildup, I avoided the areas that would receive the black and white invasion stripes.

REFERENCES

Aces W. Wayne Patton, Squadron/Signal Publications, Carrollton, Texas

P-47 Thunderbolt in Detail & Scale Bert Kinzey Squadron/Signal Publications, Carrollton, Texas

Republic P-47 Thunderbolt Martin Velek and Valerij Roman, MBI Publications

Aero Detail No. 14 P-47 Thunderbolt Art Box Co., Japan

The next chore was painting the ID bands on the horizontal stabilizers (white on top, black underneath), then the red cowl front and rudder.

Half of achieving a great finish is applying a realistically weathered look. I like to airbrush panel and hinge lines with a darker color, much like applying a dark wash.

Thinned Tamiya smoke (X-19) went over all the lines on the camouflaged portions. Tamiya medium gray (XF-20) was used over the aluminum and white areas. I airbrushed the exhaust ducts and stains behind the machine-gun shell ejector chutes with a mix of red-brown (XF-64) and flat black.

The next step was the application of a gloss coating. I used an acrylic to seal the paint beneath and protect it from a black oil wash that came afterward. The wash further enhanced the panel and hinge lines.

When the gloss was dry, I applied Gabreski's markings from AeroMaster's U.S. Top Guns sheet No. SP48-03.

Left: To seal the decals, I applied a light coating of acrylic clear flat, just enough to produce a nice satin sheen. The aluminum undersurfaces and belly tank were spared the flat coat.

The last steps were unmasking the cockpit and canopy and attaching the landing gear, propeller, machine-gun blast tubes (cut from stainless-steel tubing), pitot tube, and antennas.

I think any collection of World War II aircraft must include the plane of America's top European-theater ace. Now I've got mine!

FRANCIS "GABBY" GABRESKI

Most modelers know the name of the USAAF's highest-scoring ace in the European theater, but not so many know about the rest of this legend's flying career. The son of Polish immigrants, Francis Gabreski grew up in Oil City, Pa.

He entered the University of Notre Dame in South Bend, Ind., in 1938. After nearly flunking out his freshman year, he developed an interest in flying, and when an Army recruiter visited the campus, Gabreski signed up.

He entered flight training and nearly washed out several times, eventually getting his wings in March 1941. He miraculously landed his first-choice assignment, flying fighters in Hawaii! Gabreski got into the air on the morning of Dec. 7, but did not encounter any Japanese aircraft.

The pilot wanted to see action, and, capitalizing on his ability to speak Polish, he wrangled a transfer to one of the RAF Polish squadrons and flew Spitfire Mk.IXs. He didn't get his first Europe combat mission until January 1943. He had no aerial victories with the RAF.

In February 1943, Gabreski was assigned to Hub Zemke's 56th Fighter Group flying P-47s, and that spring he was promoted to captain and flight commander. Then in the summer, he was promoted to major and squadron commander – all this without downing a single enemy aircraft.

Gabreski shot down his first aircraft, an Fw 190, Aug. 24, 1943. The ice broken, victories came rapidly. He flew fighter sweeps over the beaches of Normandy,

and on July 5, 1944, Gabreski scored his 28th victory.

Due for a leave, he decided to sneak in one more mission before departure. While strafing a German airfield, his P-47's propeller tips struck the ground and he crash-landed. He was captured after five days of evading and lived out the rest of the war as a prisoner of war.

Gabreski continued his Air Force career after his release and became a jet ace (6½ victories) flying F-86 Sabres in Korea. He later commanded several Air Force wings.

After he retired as a colonel, Gabreski settled in Long Island, N.Y. He worked at Grumman Aerospace and served as president of the Long Island Rail Road. Francis Gabreski died Jan. 31, 2002.

– Paul Boyer

Improving the
SUPER CORSAIR

Correcting and detailing Aviation Usk's 1/72 scale F2G   BY JOE HEGEDUS

The "Super Corsair," had it gone into production, would have been the U.S. Navy's fastest climber. It was designed to combat the new Japanese threat, the Kamikaze. The end of the war brought about the cancellation of the F2G after seven pre-production examples and only 10 production machines were built.

Being a diehard Corsair fanatic, I was thrilled when Aviation Usk announced a 1/72 scale injection-molded kit of the Goodyear F2G. Though only a small number were built, it's my favorite version of the "U-Bird."

Out of the box, the Aviation Usk model is identifiable as an F2G, but there are several errors. I'll show you how I fixed them and highlight the visible detail differences between the land-based F2G-1 and the carrier-capable F2G-2. These corrections are for the Navy aircraft and won't necessarily apply to the modified civilian racers. The areas that need work are the engine, cockpit, wings, and fuselage. I know – that's just about the whole airplane.

Wings. The kit wings are incorrect, representing the metal-skinned outer panels on F4U-5 and later models, but with the correct six-machine gun armament for the F2G-2. F2Gs used wings that had fabric-covered outer wing panels, just like all Corsairs through the F4U-4. You can either add the fabric detail or use wings from another kit that has the fabric detail. I kitbashed my F2G with the wings from a Hasegawa F4U-1D kit. (The wings from the new Academy kit could be used, too.)

To modify the Hasegawa wings, first cut away the area in front of the leading edge and file off the bulges under the inner wing panels for the oil coolers, **1** and **2**. Also remove the tabs inside the upper and lower wing surfaces that position the oil cooler/intercooler detail – you'll need the clearance later. Carve the upper wing oil cooler intake lip slightly larger, **3**.

Trim the root ends of the upper wing panels, **4**. Take care not to cut all the way through the trailing-edge flap so you can keep the bottom surface. Fill the step hole in the right flap; the F2G had a spring-loaded cover over the step. (The open step is a feature unique to the F4U-4 series.) Also, sand the fabric detail off the outer wing flaps. On the F2G, these were metal-skinned just like the inner two.

The F2G used a different oil cooler configuration than the F4U; the coolers were located in the wing root rather than in the wing inner sections. Also, the F2G's R-4360 had no intercooler. The wing root intakes of the F2G were therefore smaller and shaped differently than those of standard Corsairs. I used ⅛"-long sections of ⁵⁄₃₂"

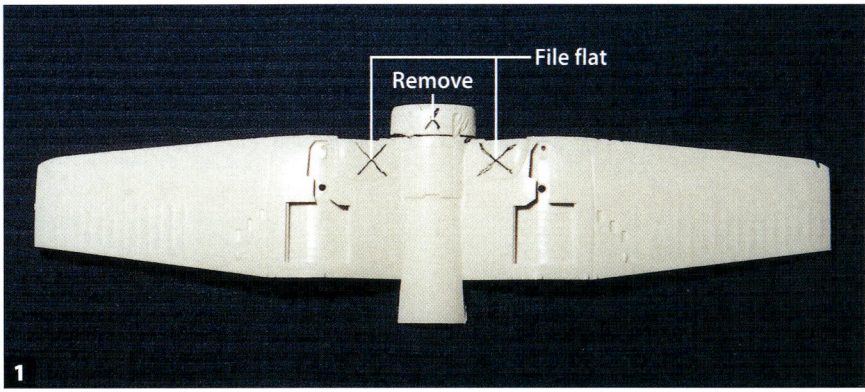

1

Remove — File flat

The underside of the Hasegawa wing showing the areas to be removed or cut down.

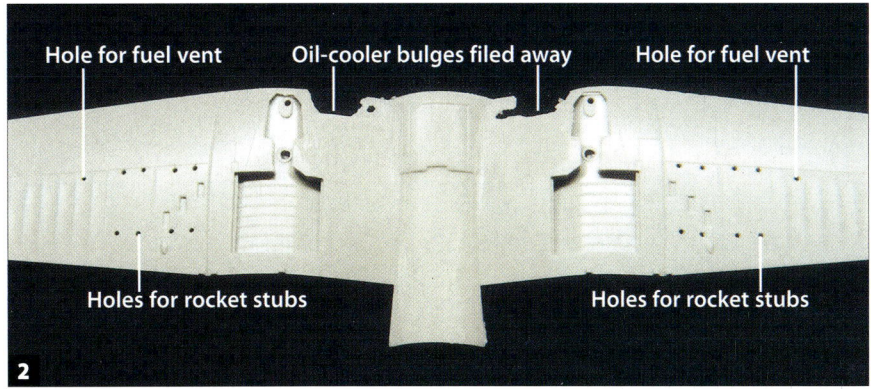

2

Hole for fuel vent — Oil-cooler bulges filed away — Hole for fuel vent

Holes for rocket stubs — Holes for rocket stubs

Holes for rocket stubs and fuel vents are drilled in the wing.

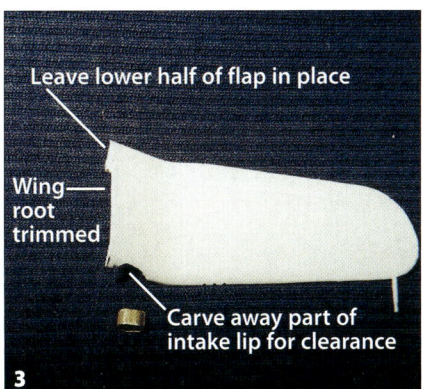

3

Leave lower half of flap in place

Wing root trimmed

Carve away part of intake lip for clearance

The oil cooler intake lip has been opened up to fit new cooler ducts. The upper wing root has also been trimmed.

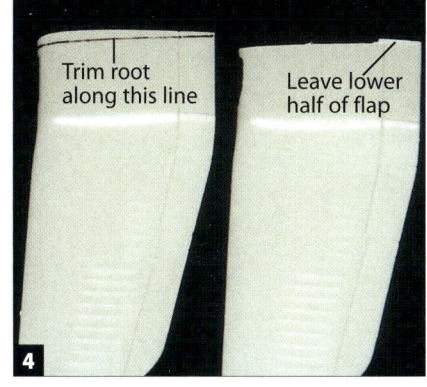

4

Trim root along this line — Leave lower half of flap

Before and after views of the Hasegawa left upper wing panels showing modifications made to the wing root.

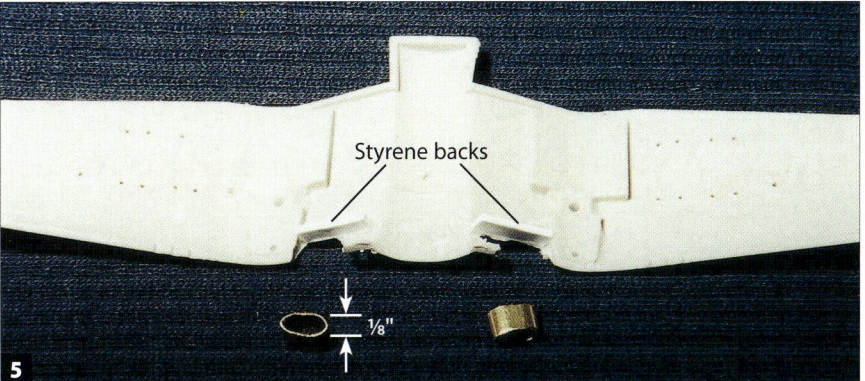

5

Styrene backs

⅛"

Sheet styrene backs the new brass-tube cooler ducts.

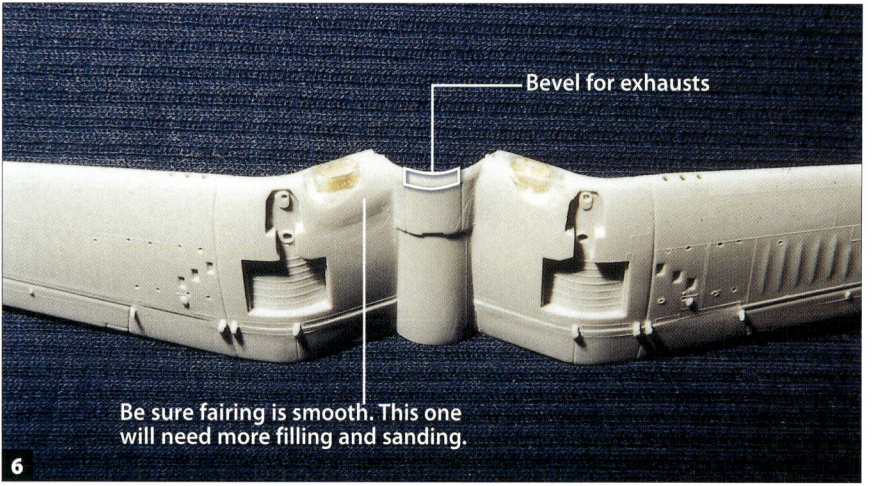

Bevel for exhausts

Be sure fairing is smooth. This one will need more filling and sanding.

6

The wing underside after the intakes have been faired in with gap-filling super glue.

½" ½"
Fuselage centerline

7

Front view of F2G-2 showing the modified oil cooler ducts. Note six-gun armament of F2G-2. The F2G-1 has two guns per wing.

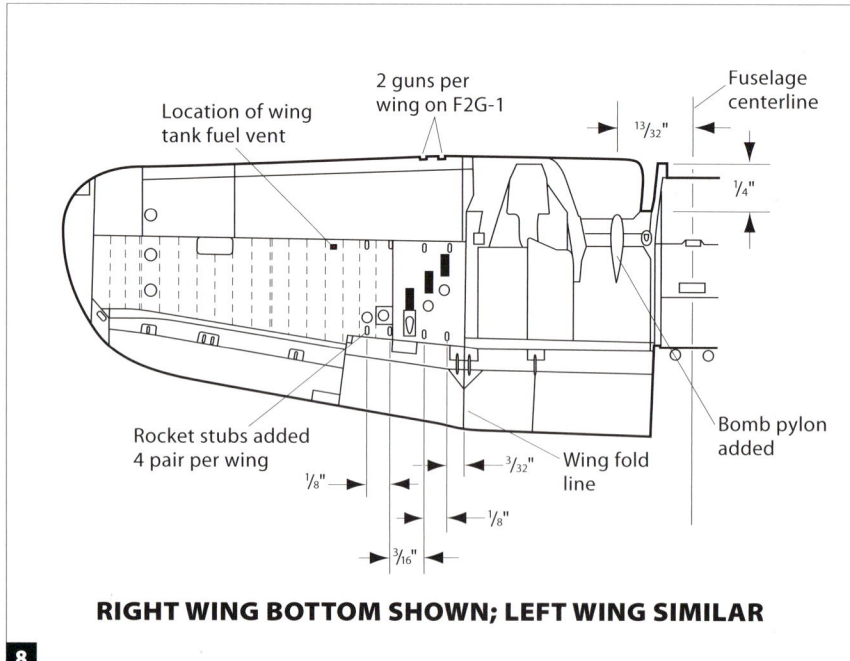

Location of wing tank fuel vent

2 guns per wing on F2G-1

Fuselage centerline

$^{13}/_{32}$"

¼"

Rocket stubs added 4 pair per wing

$^{1}/_{8}$" $^{3}/_{32}$"

Wing fold line

$^{1}/_{8}$"

$^{3}/_{16}$"

Bomb pylon added

RIGHT WING BOTTOM SHOWN; LEFT WING SIMILAR

8

F2G LOWER WING SURFACE: This shows the location of pylons, rocket stubs, and fuel vents. Check text to see which versions carried these features.

brass tube for the intakes, cleaned up the edges, and flattened them to ⅛" high, **5**.

Install $^{5}/_{32}$" x $^{7}/_{16}$" sheet styrene backs for the intakes, then assemble the wings. Slide the new intakes into the oversize openings in the wings with the inboard end ½" from the fuselage center line. Use gap-filling super glue to both secure the brass intakes to the wing and to fill the area around the intakes. When it sets up, file and sand the wing root smooth to match the rest of the wing, **6** and **7**. Use thin sheet styrene to add two vertical vanes in each intake, dividing the opening into thirds. These should have their forward edge about $^{1}/_{32}$" back from the intake leading edge. Bevel the forward edge of the center wing section to make room for the lower exhaust pipes.

If you want underwing rocket stubs on your F2G, mark the locations (and perhaps drill mounting holes) according to drawing **8**. There are no stubs in the kit, so you'll either have to scratchbuild them or find some in your spares box as I did. Several F2Gs had fuel tanks installed inside the outer-wing leading edges and had an underwing vent, **9**. Add rocket stubs and fuel vents during final assembly.

One of the big differences between the ground-based F2G-1 and the carrier-capable F2G-2 is the armament. The F2G-1 carried only four guns, the outboard guns deleted. The underwing shell-ejection ports for the missing guns may still have been there, so I didn't eliminate them. The F2G-2 carried six guns, just like standard F4U Corsairs.

Wing pylons must be scratchbuilt, as no correct ones are available. I laminated two pieces of .040" sheet styrene, then cut the pylons to shape, **10**. These will be added later in construction. Pylons were fitted to F2G-2s, but the prototypes and at least the first F2G-1 had none. When pylons were added to the F2G-1 (and at least one of the XF2G-1s), the designation became F2G-1D.

Engine and propeller. The main purpose of the F2G was to take advantage of the new 28-cylinder Pratt & Whitney R-4360 Wasp Major engine. Aviation Usk's engine represents the Pratt & Whitney 18-cylinder R-2800. The easiest way to correct this is to install a resin Engines and Things R-4360 (No. 72-003, available from Aviation Usk). This engine had to have the tops of the cylinders shaved off so it would fit in the cowl, **11**. The square-shaped protrusion on the front row of cylinders should point down when the engine is installed.

The kit propeller is usable after clean-up. The F2G-1 prop was 14' in diameter, and the kit prop is close. The F2G-2 used a smaller 13' 7" diameter prop. I used a white-metal prop taken from the High Planes 1/72 F4U-5N kit for the F2G-2, rather than trim the kit prop.

Cockpit. The F2G cockpit was almost identical to that of the F4U-4. I chose the cleaner Hasegawa cockpit for both my F2Gs. Begin by removing the rudder pedals from the floor. The stock F2G seat was just a pan with an armored backrest, **12**. Make a new backrest from .010" sheet, ¼" x 5⁄16". A pair of styrene strips represent the hinges that allow the armor plate to be moved for access to the aft fuselage.

Cut a new rear bulkhead/armored headrest from .020" styrene, and add that to the cockpit tub. The kit's armor headrest is undersize and positions the cushion too low relative to the seat. Carve a new cushion from sprue or cut off the kit cushion and mount it on the new bulkhead.

Attach the rear bulkhead to the floor and the rear ends of the consoles. After painting the cockpit tub, add the console decals, glue the seat to the floor, and then add the armored backrest to the seat and bulkhead. I added belts and harnesses from lead foil, but the kit photoetched harnesses could be used. The Hasegawa stick was installed.

The kit's photoetched instrument panel looks great, but the layout isn't correct for the F2G. I used it as a template to cut a new panel from sheet styrene. The gauges came from ProModeler's 1/72 scale U.S. Navy WWII fighter instrument panel decal sheet (No. 88101100200) for the F4U-4. I attached the photoetched rudder pedals to the back of the instrument panel. The cockpit was painted interior green below the consoles, black above, **13**.

Fuselage. The fuselage work is pretty straightforward. Carefully open the exhaust troughs under the cowl flaps with a sharp knife and files, and grind away the mold stubs from inside the cowling. You'll also need to extend the cockpit opening about 1⁄32" at the rear to allow the new armored headrest to fit.

Assemble the fuselage with the engine, cockpit, and tailwheel-well roof installed.

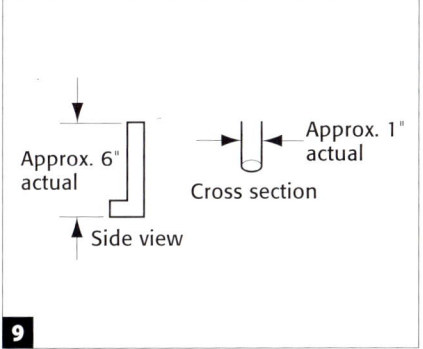

9

FUEL VENT: Dimensions are for the real item. Make from stretched sprue.

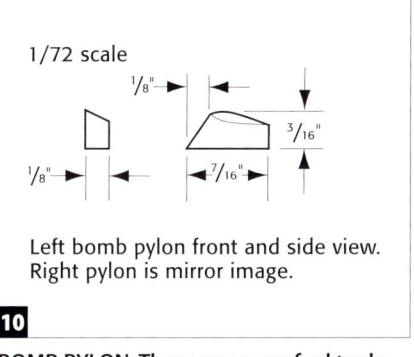

10

BOMB PYLON: These can carry fuel tanks or bombs under the inner wing sections.

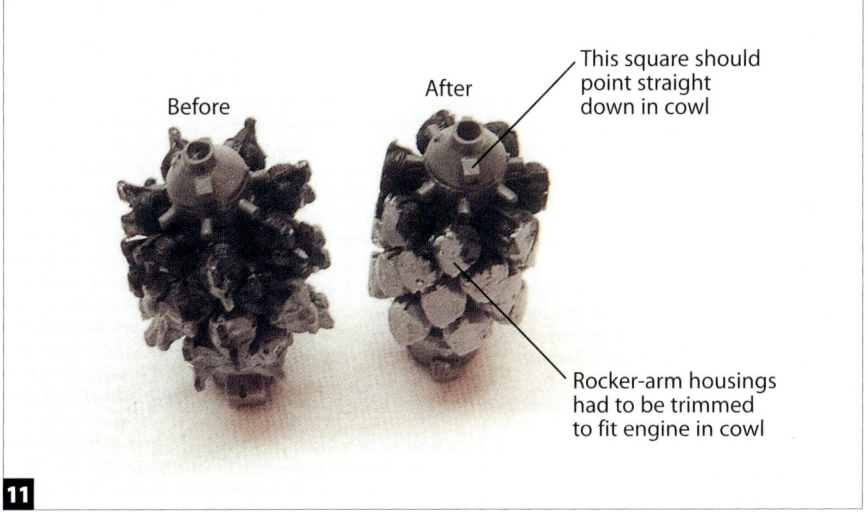

11

It's easy to see why the Pratt & Whitney R-4360 Wasp Major engine was nicknamed the "corncob." Joe had to shave the tops of the Engines and Things resin engines so they would fit inside the cowl.

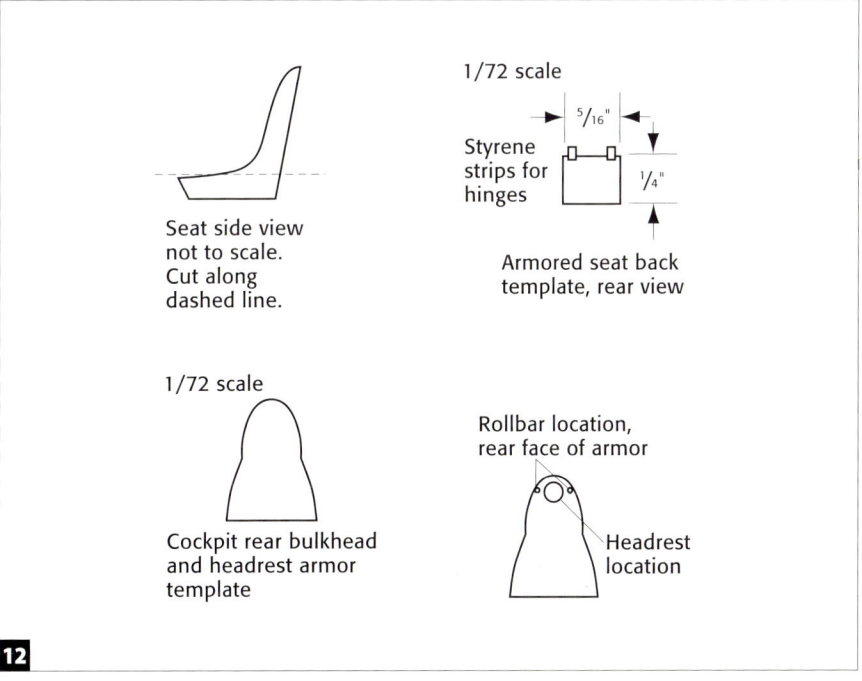

12

COCKPIT DETAILS: A new rear bulkhead, armor plate, and seat need to be made for the cockpit.

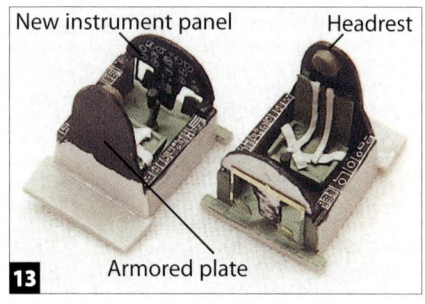

New instrument panel   Headrest

13

Armored plate

Fore and aft views of the completed cockpit tubs.

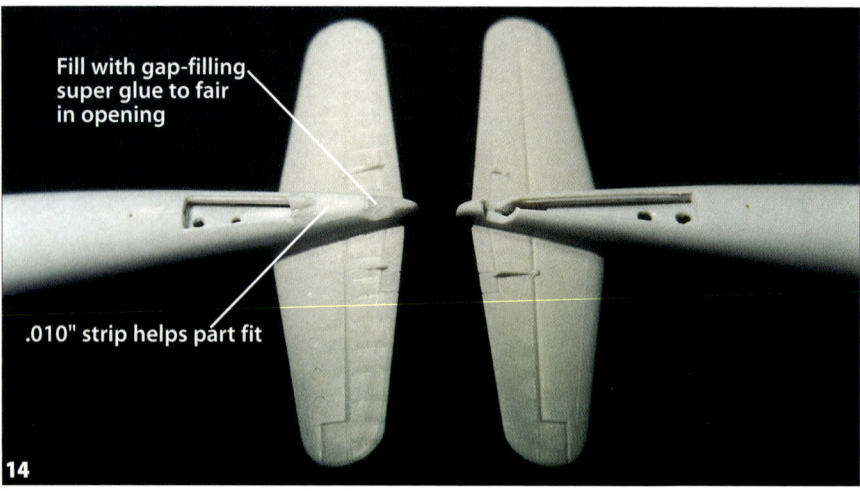

Fill with gap-filling super glue to fair in opening

.010" strip helps part fit

14

The F2G-1 (left) has the aft part of the well faired over. The F2G-2 (right) had a tail hook with the full bay.

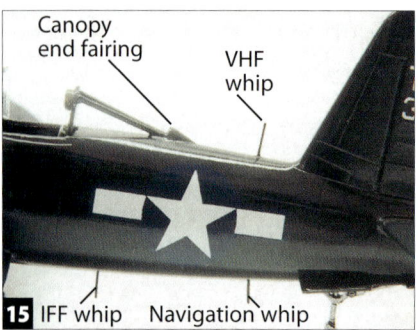

Canopy end fairing   VHF whip

15 IFF whip   Navigation whip

The aft fuselage of the F2G-2 shows the canopy end fairing, rollover bars, and whip antennas.

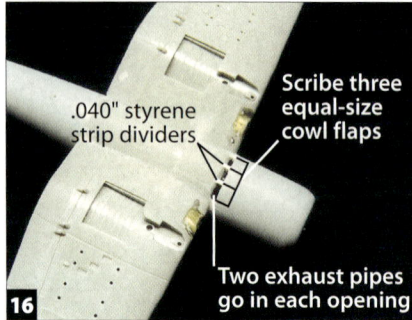

.040" styrene strip dividers

Scribe three equal-size cowl flaps

Two exhaust pipes go in each opening

16

This underside view shows the three cowl flaps and dividers for lower exhaust troughs.

Make sure the engine is centered in the cowl and the cylinders are aligned properly. I secured the engine with super glue around the aft cylinder bank, being careful not to use too much. You don't want to plug the exhaust troughs.

On the underside of the fuselage, remove the outer cowl flap from each side and fill the scribed lines between the remaining cowl flaps. The F2G had only three cowl flaps underneath rather than the four that remain

after you've removed the outer ones. Scribe new cowl-flap divisions, so you have three equal-size flaps on the bottom.

The tailwheel-well roof will need trimming at its aft end to fit between the fuselage halves. Also, to model the F2G-2, you'll need to enlarge the rear of the opening to allow clearance for the hook.

The F2G-1 did not have a tailhook, so the rear of the opening was faired over. The kit fairing is too short, but can be used with modifications. Place a strip of .010" sheet styrene under the kit part and fill the hook cutout with gap-filling super glue so that the aft fuselage is solid, **14**. The F2G-2 tailwheel well is the same as that of an F4U-1D.

If you want to pose the canopy open, remove the point from the rear of the kit canopy. This is actually part of the fuselage, and the canopy opens over it. You can either add the canopy point to the fuselage or carve a new fairing from plastic, **15**.

The rollover bars go from the fuselage just in front of this fairing to either side of the armor headrest (see **12**), forming a V with the point to the rear. I made new bars from brass tube, but the kit rollover bars could be used. Separate the windscreen from the canopy, and add those to the appropriate positions during final assembly.

Main assembly. Join the fuselage and wings, and fill and sand any wingroot gaps. Add the horizontal stabilizers from the Hasegawa kit after scribing five circular access panels on the upper left and lower right surfaces. I sharpened the end of a piece of 1/16" brass tube and pressed and twisted it into the surface to produce the circles. Align two .040" strip dividers with the scribed lines dividing the bottom cowl flaps, **16**. Now add the wing pylons as shown in drawing **8**.

Landing gear. I used Hasegawa's landing gear with True Details resin wheels, both improvements over the Aviation Usk parts. I extended the tail wheel strut 1/8" by replacing the kit portion with a piece of stainless steel tubing. If modeling the F2G-1, clip off the tailhook.

Paint and markings. My models represent an F2G-1 and an F2G-2 that were assigned to the Tactical Test Division of the Naval Air Test Center in 1946. Both are

REFERENCES

F4U Corsair in Detail and Scale (Parts 1 and 2) Bert Kinzey, Squadron/Signal Publications, Carrollton, Texas, 1998

F4U Corsair In Action Numbers 29 and 145; Jim Sullivan, Squadron/Signal Publications, 1977 and 1994

F4U Corsair Warbird History Nicholas A. Veronico, John M. Campbell, and Donna Campbell, Motorbooks International, Osceola, Wisconsin, 1994

The Official Monogram US Navy & Marine Corps Aircraft Color Guide, Vol. 2 John M. Elliott, Monogram Aviation Publications, Sturbridge, Massachusetts, 1989

I Flew Them First Armstrong, Champlin Fighter Museum Press, Mesa, Arizona, 1994

overall Glossy Sea Blue, but the F2G-2 has a yellow-and-blue checkerboard pattern on the cowl. I used Testor Model Master enamels, with Chrome Yellow and Blue Angels Blue on the cowl. I painted the cowl yellow first, and after it was dry, I cut tiny squares and wedges (for the front of the cowl) from masking tape, laid out the checkerboard pattern, and sprayed the blue. The landing gear struts for these particular F2G's were aluminum, and the instrument coamings flat black.

After painting, I applied two coats of Future floor polish, then applied the decals. The white letters and numbers are dry transfers applied to clear decal film and coated with Microscale Liquid Decal Film, then applied as a normal decal. The national insignia are from AeroMaster, but the kit decals could be used instead. A coat of Future over the decals sealed them and gave a uniform gloss to the airplanes.

Final assembly. Add the small parts – rocket stubs, fuel vents, and stores of your choice. I hung a drop tank from a 1/72 scale Monogram F8F Bearcat on the F2G-2, and Tiny Tim rockets made from 1/32 scale Hasegawa F6F Hellcat rockets on the F2G-1. They needed to be shortened ⅛" just in front of the fins to be the correct length for a 1/72 Tiny Tim.

Make a set of exhaust pipes from ¹⁄₃₂" brass tube, rod, or stretched sprue. A pair of pipes goes in each side and bottom trough, totaling 14 pipes.

I added a scrap plastic gunsight and gun switch boxes to the top of each instrument coaming. The forward antenna masts came from the Hasegawa kits, and whip antennas from stretched sprue (see **16**). Add the antenna wire from stretched sprue or monofiliment, then set your F2G on the shelf and admire it!

THE GOODYEAR F2G

The first production F2G-1 at the Naval Air Test Center at Patuxent River, Md., in 1947. U.S. Navy photo via National Museum of Naval Aviation and Bert Kinzey

In the late years of World War II, the Japanese military developed its Kamikaze corps, a group of dedicated pilots who would guide their bomb-laden aircraft directly into a target.

It turned out that the Kamikazes didn't make much difference in the course of the war, but their actions led to several U.S. attempts to eliminate the suicide planes before they could damage critical fleet assets. One of the efforts involved a fighter that could climb to combat altitude quickly, overtake the suicide planes with speed, and overwhelm them with firepower.

The U.S. Navy saw potential for just such an aircraft in the F4U Corsair. Powered by the new 28-cylinder R-4360 "Wasp Major" engine and featuring a cut-down fuselage spine and bubble canopy, the radically redesigned F2G Corsair proved to be just the right air-

craft for the job. But the job changed before it could enter service.

The Navy had asked that the "Super Corsair" be ready for production in the spring of 1945, and an order was placed for the production of 418 F2Gs. After solving a few teething troubles with the seven XF2Gs, the Goodyear was ready for mass production.

The impending defeat of Japan and the advent of jet fighters spelled doom for the Super Corsair, however. When the order was canceled on May 8, 1945, only the 10 aircraft on the production line at the time were finished. Five were set up as ground-based F2G-1s, and five as carrier-capable F2G-2s. A couple of the aircraft were sold to air racers, and the first production F2G-1 was preserved and restored for display at the Champlin Fighter Museum in Mesa, Ariz.

– *Paul Boyer*

Both the carrier-capable F2G-2 (left) and the ground-based F2G-1 can be built from modified Aviation Usk kits.

PROJECT 4

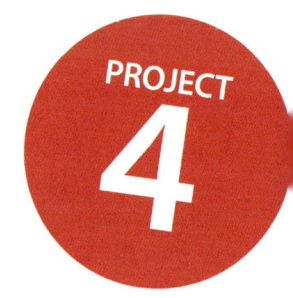

Finishing O'Hare's WILDCAT

Painting and weathering Revell's classic 1/32 scale F4F BY MIKE ASHEY PHOTOS BY GLENN JOHNSON

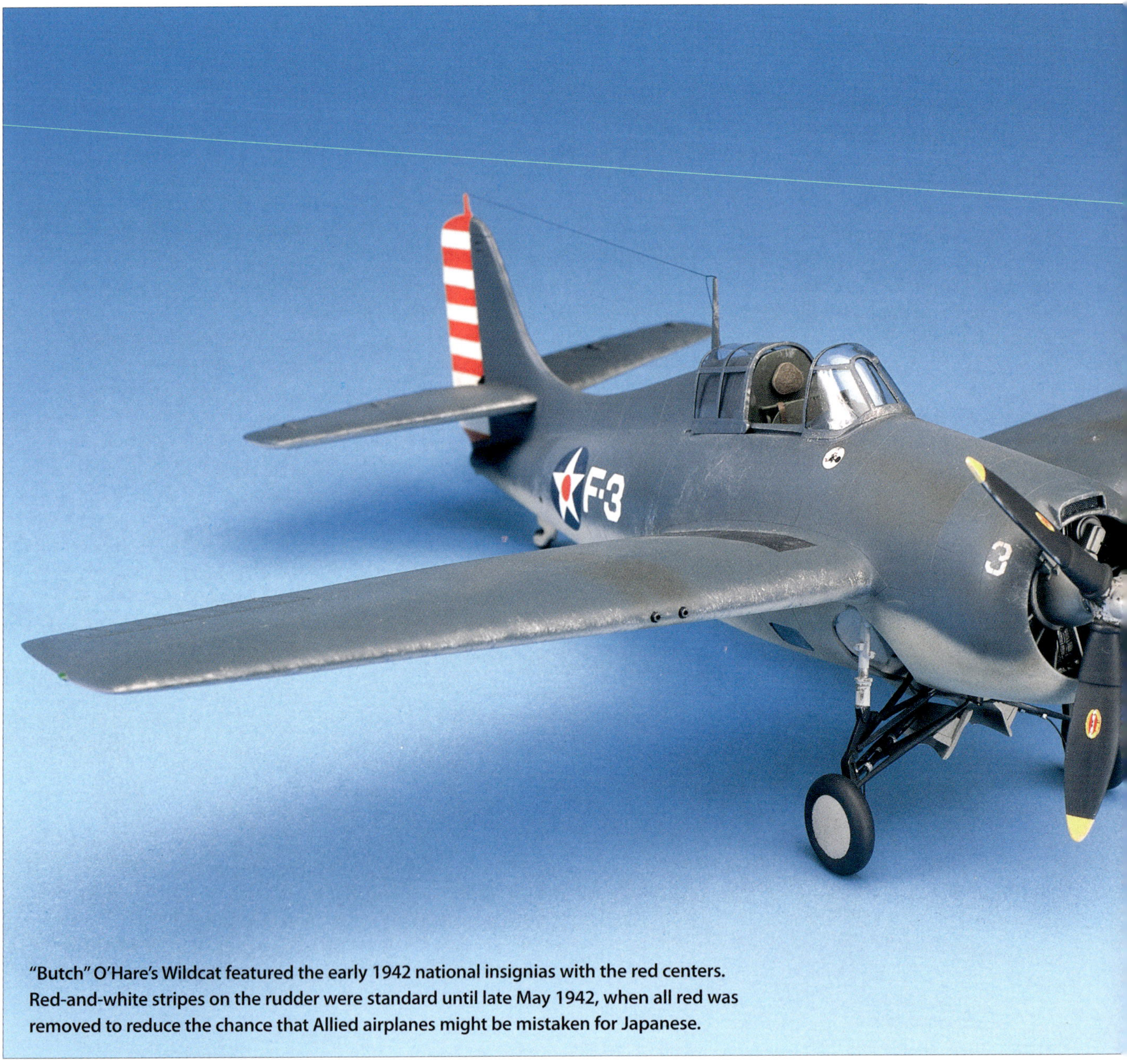

"Butch" O'Hare's Wildcat featured the early 1942 national insignias with the red centers. Red-and-white stripes on the rudder were standard until late May 1942, when all red was removed to reduce the chance that Allied airplanes might be mistaken for Japanese.

I like building World War II Pacific Theater U.S. Navy aircraft, and I wanted to have a 1/32 scale Wildcat in my collection. Revell's kit has been around since the early 1970s, and although not great, it's the only 1/32 scale Wildcat available.

I wanted to model the aircraft flown by Edward "Butch" O'Hare from the USS *Lexington* when he single-handedly shot down five Japanese G4M "Betty" bombers on Feb. 20, 1942, earning him the Medal of Honor. He was killed in November 1943, possibly by friendly fire. Chicago's O'Hare International Airport is named for him.

I won't have enough space here to show you all I did to improve the old Revell classic, but if you want to see some of my construction work, check out my book *Model Aircraft Tips and Techniques, an Illustrated Guide* (Kalmbach). In this article, I'll concentrate on painting and weathering the model.

For references, I used Bert Kinzey's *F4F Wildcat in Detail* (Squadron/Signal), and the beautiful Watanabe illustration in the Hellcat chapter of *The Great Book of World War II Airplanes* (Bonanza).

Overall finishing. I used three finishing techniques after color application. First, I applied a wash of a dark complementary color to visually deepen recesses. Second, I dry-brushed a lighter version of the base color onto high points to simulate wear and make relief moldings more visible. A light touch is all that is needed. Third, I scraped artist's pastel pencils to form little piles of colored dust and brushed them onto the model to simulate exhaust and gunpowder stains, fluid leaks, and grime. A light coat of clear flat sealed the pastels.

There are many ways to paint models, and you can use several techniques on the same model. Don't settle for simply painting, when you can wash, dry-brush, and apply pastel dust for that extra touch of realism.

I primed all the interior subassemblies, since many of them had styrene strip and rod additions. I used a flat gray Testor enamel for the primer.

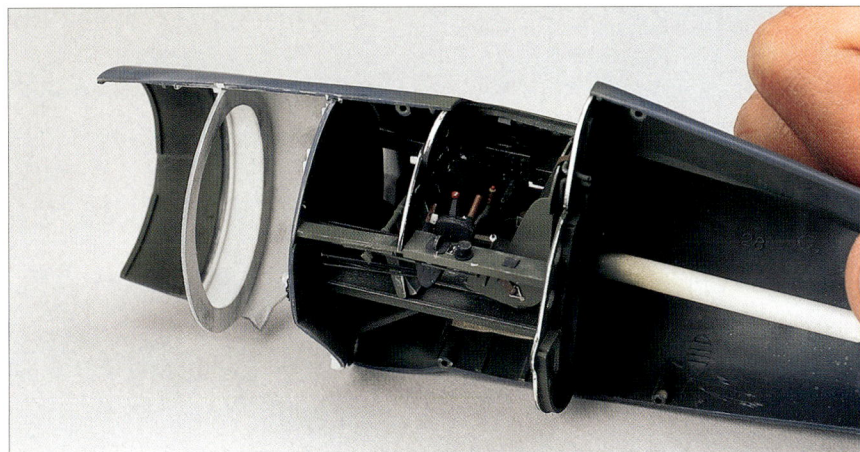

I painted the cockpit and cowl interiors with interior green, then the wheel-well area flat white. Details in the cockpit came next, accented with flat black mixed with a little white to form a charcoal gray. I attached Waldron instruments to the panel. The seat harness was made from masking tape, painted with Testor "wood" enamel, and dry-brushed with white. Note the Wildcat had no floor – just a pair of foot troughs leading up to the rudder pedals.

The wheel-well bulkhead framing and bolt heads were dry-brushed with Testor silver. I created the grimy look with dark washes and dark gray pastel pencil dust. I painted the cockpit in two shades of interior green to add realism. Black pastel dust was applied to the recesses and all the edges were dry-brushed with Testor silver to simulate wear.

After painting the engine cylinder banks gloss sea gray, I applied Testor Metalizer burnt metal as a wash to make the cylinders' cooling gills stand out. Note the photoetched screening for the engine air intake and the wiring added to the engine.

I airbrushed silver paint over putty-filled areas to make it easier to see remaining flaws. Before painting, though, I removed the silver, since other paints don't stick well to it. I also cleaned the model with Polly S Plastic Prep, which removed sanding dust and fingerprint oils that may inhibit paint adhesion.

The landing gear and the engine area were carefully masked with small strips of masking tape. I primed and checked once again for flaws, then filled them with gap-filling super glue and sanded them smooth. The light spots are filled flaws.

Masking a cockpit area can be tedious, and you have to be careful not to break off small interior parts. Small wads of facial tissue were stuffed into the cockpit and then covered with small strips of masking tape. Several layers of tape may be needed. After the second priming, the model was ready for final colors.

After masking the tail area, I painted the rudder with several coats of flat white. White can be hard to paint. Flat white covers better than gloss, and it won't build up too quickly if you apply light coats.

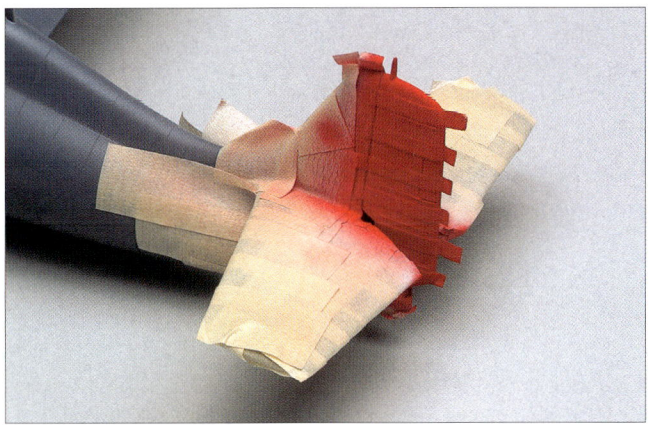

I let the flat white dry for a few days, then applied strips of masking tape (cut with a brand-new No. 11 blade for extra-sharp lines) to cover the white rudder stripes. I airbrushed flat red next, and when the paint was dry to the touch, I removed the tape strips. The painted rudder dried for another couple of days, then it was masked again.

The light gray color was next, covering the entire bottom of the wing, fuselage, and stabilizers. The bottom color was called "light gray" and was equivalent to today's flat light gull gray.

THE GRUMMAN WILDCAT

In the mid-1930s, new powerful aircraft engines were making larger and faster aircraft possible. Engineers could see this progress would be hindered by biplane aerodynamics – two wings and many wires create a lot of drag.

After a trio of naval biplane fighters (FF, F2F, and F3F), Grumman designers came up with a monoplane follow-on. The first XF4F-2 flew Sept. 2, 1937. At first, the U.S. Navy wasn't pleased with the new monoplane because of engine performance and favored the competing Brewster XF2A-1 Buffalo.

Eventually the Wildcat was put into production as the F4F-3. Orders from the Navy and from France resulted in the construction of 380 F4F-3s. The Wildcats destined for France were eventually issued to Great Britain as Martlets, and some, designated F4F-3A, were assigned to the U.S. Navy.

An improved Wildcat, with six (instead of four) machine guns and foldable wings followed. The 1,169 F4F-4s saw service with the U.S. Navy and Britain's Fleet Air Arm as Martlet IV.

Grumman decided to transfer Wildcat (and Avenger) production over to General Motors and concentrate on building the new F6F Hellcat. GM produced 839 of its four-gun version of the F4F-4 as the FM-1 (Martlet/Wildcat V).

The ultimate Wildcat was the FM-2 (Martlet/ Wildcat VI). It had a taller tail, a nine-cylinder Wright R-1820 Cyclone engine (as opposed to the previous 14-cylinder Pratt & Whitney R-1830 Twin Wasp), and other minor changes.

The FM-2 was the most-produced version of the Wildcat, with 4,737 built by GM. It saw service in both Atlantic and Pacific theaters, flying mostly from the smaller escort carriers.

The last FM-2 was built in May 1945, so the Wildcat was unique among U.S. Navy carrier-borne fighters as it saw production and service all the way through World War II.

– Paul Boyer

I airbrushed the top-side color blue-gray, mixed from intermediate blue and gloss sea blue. This color is now available from Testor in enamel and Polly Scale in acrylic. The color looks too dark here, but it was lightened with a coat of clear flat after decaling.

The entire model received two coats of clear gloss. I used polyurethane on this model, but any clear gloss could be used. The glossy surface allows the proper adhesion of decals and helps prevent "silvering" caused by air trapped beneath decals. Cutting away extra clear film and using a decal setting solution work, too, making the decals conform tightly to both raised and recessed surface detail.

After a clear flat coat, weathering was next. I lightly airbrushed flat black smoke stains from the machine guns and the shell ejector chutes in the direction of the airflow. Building up with several light passes allowed for adjustments to be made. I made the engine-exhaust stains by brushing pastel dust (a mix of black and brown) away from the exhaust pipes. The dust will cling to the flat finish, but another coat of clear flat will seal it.

I dry-brushed silver paint onto the edges of the cowling, leading edges of the wings, stabilizers and fin, wing walks, and around the canopy. To dry-brush, dip a brush into silver paint, then wipe most of the paint onto a clean rag. Whisk the brush on high points and leading edges and the remaining pigment on the brush will create a realistic worn-paint appearance.

I also dry-brushed the windscreen frame and cockpit sills with silver paint. These areas wear from handling and foot traffic. You can get carried away with dry-brushing, so go slowly. Note the scuffing on the wing walks.

REFERENCES

Wildcat Aces of World War 2 Barrett Tillman, Osprey Publishing, London, England, 1995

F4F Wildcat in Detail Bert Kinzey, Squadron/Signal Publications, Carrollton, Texas, 2000

F4F Wildcat in Action Don Linn, Squadron/Signal Publications, Carrollton, Texas, 1988

Below: The classic plank wing planform of the Wildcat is most evident from behind. The F4F-3 Wildcat did not have a folding wing; that device came along with the F4F-4.

Detailing
MINSI II

Improving Hasegawa's big 1/32 scale Hellcat   BY MIKE ASHEY   PHOTOS BY MIKE ASHEY AND GLENN JOHNSON

Hasegawa's 1/32 scale Hellcat has been around a long time, and if you like building in this scale, it's the only game in town. I decided to build a Hellcat to complete my collection of U.S. Navy Pacific war fighters and to complement my Revell 1/32 scale Wildcat and Corsair.

Hellcat subjects can be diverse, but I wanted to do one of the aircraft flown by the Navy's top ace, Commander David McCampbell. The kit came with markings

for his "Minsi III," but the decals were not correct, so I obtained an EagleStrike decal sheet for McCampbell's earlier "Minsi II."

To dress up the interior, I used a Verlinden Productions resin interior set along with an Eduard photoetched brass detail set. The combination was necessary, as neither set included every detail I wanted to add.

I made other improvements along the way. For reference material I used Bert Kinzey's *F6F-3/5 Hellcat in Detail and Scale*

and *The Great Book of World War II Airplanes*, published in the mid-1980s by Bonanza Books. In addition, I took a number of photos of actual Hellcats at the National Museum of Naval Aviation in Pensacola, Fla.

Although the kit took longer than I expected to build (my wife dubbed it "the never-ending model"), it looks great next to my Butch O'Hare Wildcat and Pappy Boyington Corsair.

Test assembling the kit with tape reveals fit problems and gives you an opportunity to come up with solutions before applying glue.

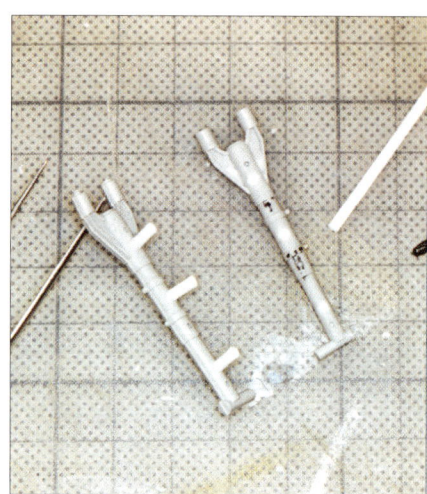

Ejector-pin marks spoiled the landing-gear struts. I drilled them even deeper, glued in sections of plastic rod, then cut and sanded them flush with the struts.

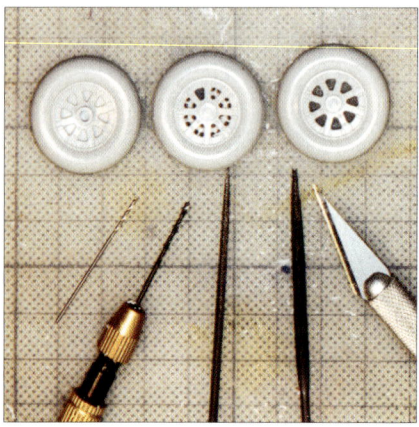

To open the wheel spokes, I first drilled three holes in each opening, then cut the remaining plastic away with a No. 11 blade and finished with needle files.

After gluing the halves together, I separated the elevators from the horizontal stabilizers by scoring along the hinge lines with a panel scriber and cut through with a No. 11 blade.

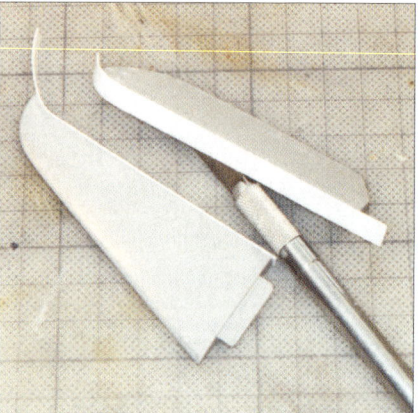

I filled the stabilizers and elevators with liquid resin, then covered the openings with strip styrene attached with super glue. The covers were then shaped with a blade and sanding sticks.

I taped the elevator to the stabilizer and then marked the hinge tab positions. After drilling out the tab locations, I inserted .010" x .060" styrene strips and superglued them in place.

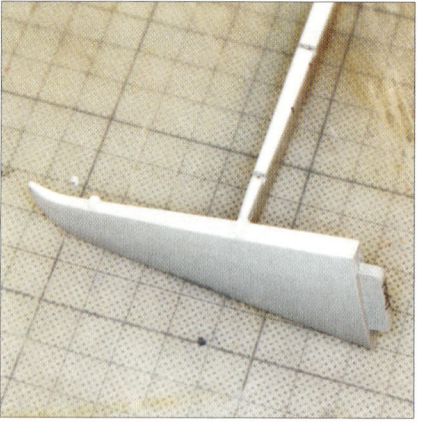

When the strips were set, I removed the elevator and reinforced the strip tabs with super glue. I then cut and sanded the tabs to shape.

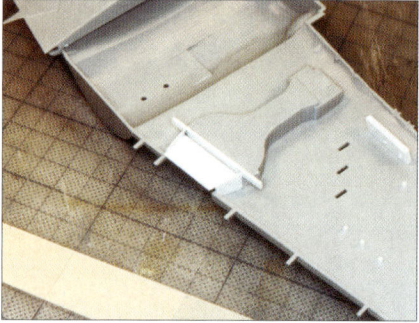

I built up the rear walls of the main gear wells with sheet styrene. The strip behind the machine-gun ports was drilled out to hold thin plastic-tubing gun barrels that will be inserted after the model is painted. I also opened the shell-ejection ports and filled the rocket-launcher-stub mounting holes with plastic rod.

Here's how the finished wheel well looks from the outside. Note the over-size recessed panel lines. These will be replaced soon.

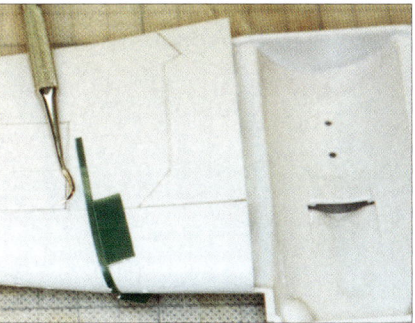

All the panel lines were filled with super glue and sanded smooth. I made new panel lines with two passes of a Bare-Metal Foil engraver. I used labeling tape as a guide tool. Cut narrow, the tape can bend a bit; wide pieces placed alongside keep it from moving. I inked each line to provide a visual reference for additional panel lines.

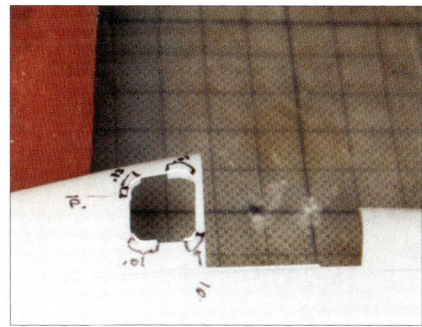

After cutting out the thin plastic ports for the small rear-view windows, I found they didn't fit well. I dry-fitted the windows, shimmed the areas with .010" and .015" styrene strip, and faired them in with fine files. I also removed the windows' mounting tabs and filled the indentations inside the fuselage with styrene half-discs.

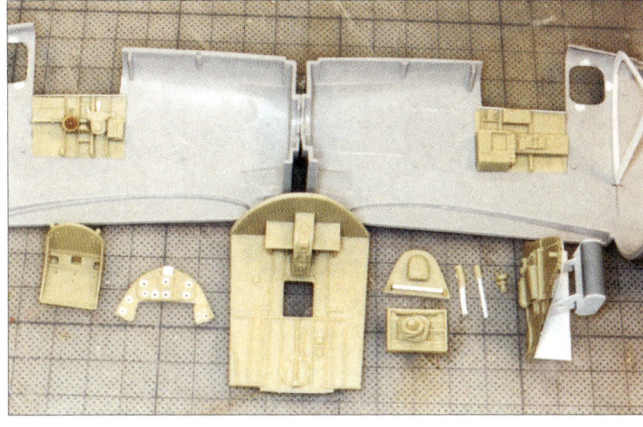

Here are the major Verlinden pieces after removing them from their resin pour stubs. Some of the items had to be replaced with styrene rod and sheet. The Verlinden cockpit details had to be test-fitted and trimmed to fit correctly. I drilled out the instrument panel to match the Waldron instrument sizes and glued a .020" sheet to the back of the console to help hold the printed instruments.

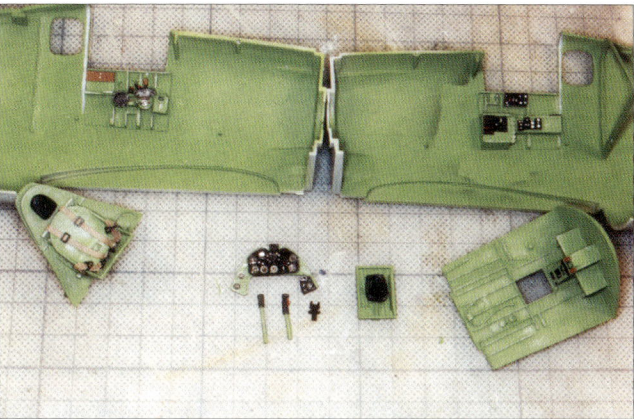

I dry-brushed silver over the interior green panels, consoles, and seat. The Waldron printed metal instruments were set in the Verlinden panel. Eduard's buckles were placed on masking-tape seatbelts. I ended up removing the scratchbuilt tank from behind the seat bulkhead because it didn't fit and couldn't be seen anyway.

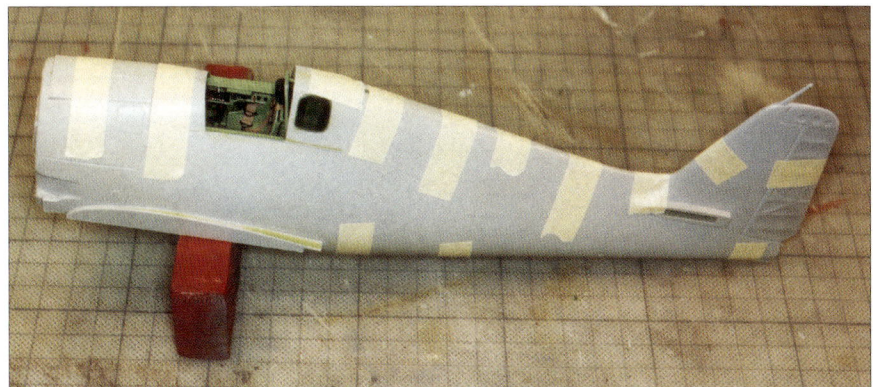

Before closing the fuselage, I inserted the small rear windows. This was tricky since I had cut off their mounting tabs. By taping over the openings on the outside of each fuselage half, I inserted the windows from the inside. Using tiny drops of gap-filling super glue applied with a thin wire, I anchored the windows in place.

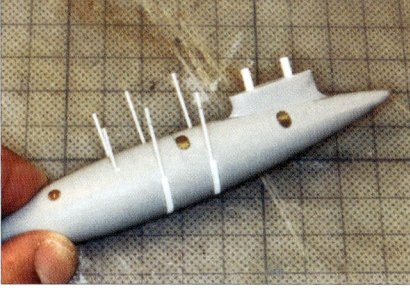

The braces for the kit's drop tank were oversize, so I replaced them with strip styrene. I also replaced the short mounting pegs with longer ones of styrene rod. All the attachments were cut long and fitted into holes drilled in the fuselage. Note the photoetched brass caps from Eduard.

The kit's push rods didn't line up with the cylinders so I cut them off and replaced them with .025" rod. I also added bits of .035" rod to the wiring harness and then drilled out their ends with a No. 72 drill bit to hold 34-gauge brass beading wire for the ignition leads. The kit's exhaust pipes didn't fit well behind the engine, so I cut off most of them and placed the drilled-out ends in the cutouts and behind the cowl flaps.

After assembling the major components, filling and sanding seams, and masking the cockpit, I airbrushed an overall coat of flat light gray primer to help reveal imperfections. Next was a bit of flat white at the top of the fin and behind the wing for the CAG band on the fin and the skinny foot guides to the cockpit. With thin strips of masking tape over these stripe areas, I airbrushed several coats of Testor Model Master gloss sea blue lightened slightly with gloss white.

After waiting a week for the gloss sea blue enamel to dry, I airbrushed a coat of interior green for the main wheel wells. It's possible that the wells of Hellcats were painted gloss sea blue, but I like the contrasting colors. When the green was dry, I dry-brushed with silver and sprinkled in dark pastel chalk dust to simulate grime. A coat of clear flat sealed the dust.

To keep the possibility of decal silvering to a minimum, I trimmed as much of the clear film from each decal as possible. This meant cutting the large insignias apart. I applied the star portions first to make it easier to align the bars.

The Eagle Strike decals went down perfectly, and I sealed them to the surface with a thinned coat of clear gloss polyurethane. When that was dry, I added the painted canopy and windshield. The windshield didn't fit well, and I had to use thin sheet styrene and white glue to fill the gaps at the bottom edges. The last steps were adding the painted engine, cowl, prop, and landing gear.

REFERENCES

F6F Hellcat in Detail and Scale Bert Kinzey, Squadron/Signal Publications, Carrollton, Texas, 1996

Great Book of World War II Airplanes Jeffrey Ethell, Bonanza Books, New York, 1984

PROJECT 6

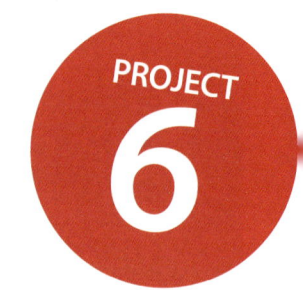

D-Day SPITFIRE

Converting Hasegawa's Mk.V Spitfire to a Griffon-engined Mk.XIV BY JEFF HERNE

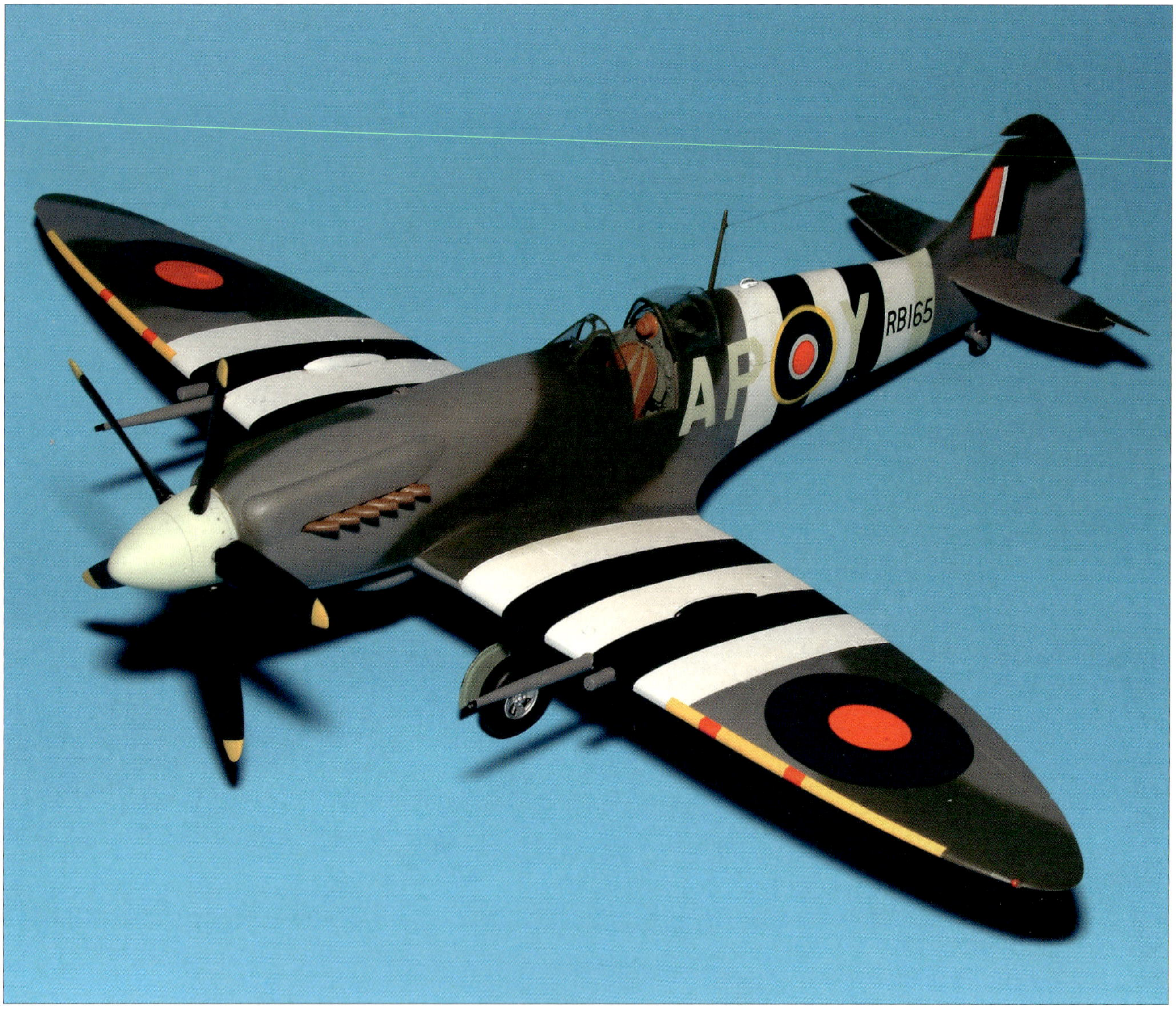

Between the D-Day stripes, yellow leading-edge bands, and colorful insignia, it's hard to believe this aircraft is camouflaged.

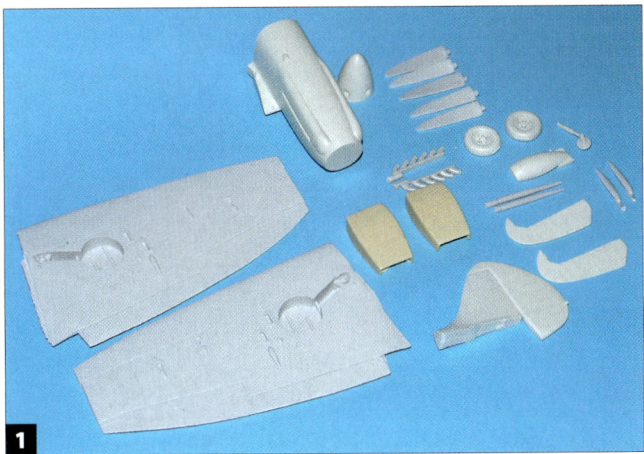

1 The Warbird Productions conversion kit includes a new nose, tail prop, and spinner. The universal "C-wings" are a separate part and aren't included in the conversion kit.

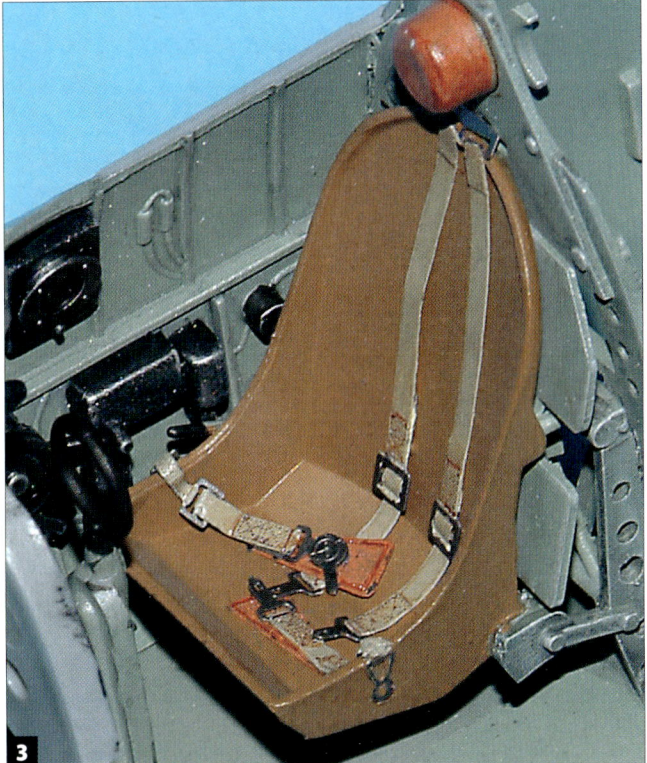

2 The Spitfire's cockpit is well detailed for a kit that's more than 30 years old. Jeff painted the instrument panel black, dry-brushed the instruments with silver, then painted the instrument faces with clear gloss.

3 A set of photoetched RAF Sutton harnesses was all that was added to the otherwise stock cockpit.

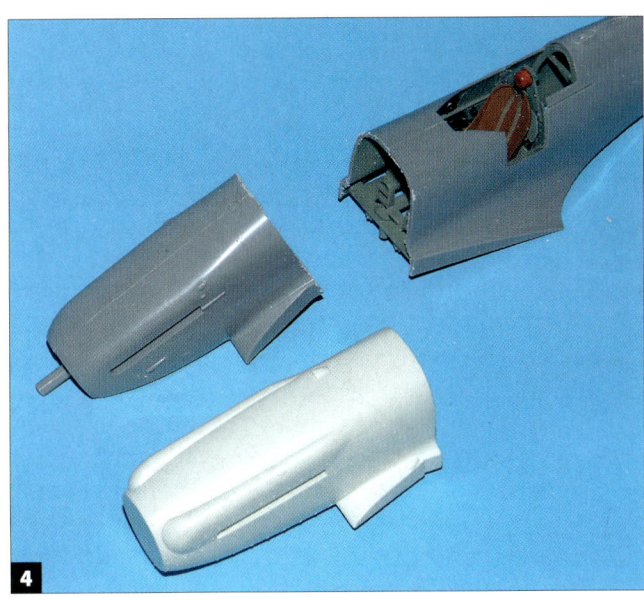

4 Using the wing root as a gauge, Jeff cut the forward fuselage with a scribing tool and razor saw to remove the Merlin cowling and make room for the bigger Griffon.

Spitfire enthusiasts constantly argue about which variant of the venerable fighter was the prettiest of all. Some believe the original Mk.I from the Battle of Britain, others say the Mk.V or Mk.IX, and some insist the late-war Griffon-engined variants are the most stunning of all. While I readily admit it's a beautiful airplane in all its variants, I'm drawn to the late-war Mk.XIV with the five-bladed Rotol prop, large spinner, and extended nose. It's a powerful-looking fighter that retains the grace and curves of the earlier versions.

Unfortunately, a complete kit the Mk.XIV has never been produced in my favorite 1/32 scale, so a conversion kit was my only option. The basis of my conversion is the Hasegawa 1/32 scale Mk.V. The Mk.XIV shared the same fuselage as the Mk.V but used a different wing and tail and, of course, a new engine. I was fortunate enough to find a Warbird Productions conversion kit and replacement wings at a local model show, **1**. That was all the inspiration I needed to jump into this project.

A cut above the rest. Hasegawa's Spitfire Mk.V (kit No. 08052) is the only Mk.V available in 1/32 scale, and the fit and detail level are quite good. I started by building the cockpit and painting the fuselage interior, **2**. The kit's cockpit was detailed enough for my liking, so all I added was a set of photoetched seat belts, **3**.

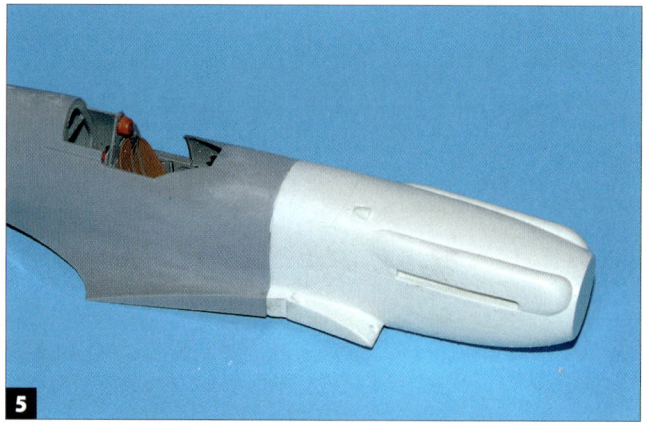

5 After marking the inside of the fuselage, Jeff added square styrene stock to strengthen the joint …

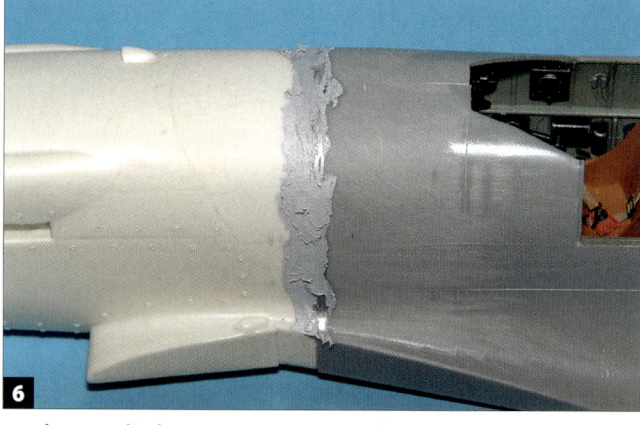

6 … then applied a generous amount of gap-filling super glue and putty to blend the joint.

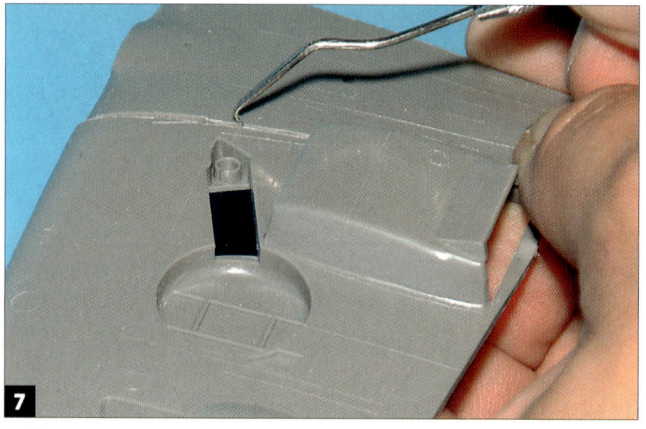

7 Jeff scribed the panel lines along the lower wing to remove the center section.

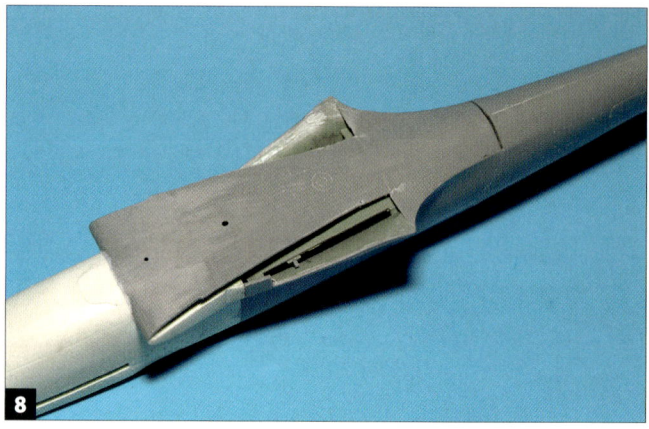

8 The center section of the wing forms the strongpoint for the outer wing sections. Jeff packed the area between the cockpit floor and center section with epoxy putty.

I assembled the fuselage and glued it together with liquid cement, then I removed the front part of the fuselage with a scribing tool and razor saw, **4**. I test-fitted the resin nose to the fuselage and marked the inside of the fuselage with a pencil. I glued styrene stock along the line I marked to create a stepped joint and glued the nose into place, **5**. The styrene provided some additional support to the joint. I filled the seam with putty, **6**, and sanded the area smooth.

On a wing and a prayer. The Mk.XIV Spitfire utilized a "C-type" universal wing, different from the Mk.V. The kit's lower wing section, including the center section of the fuselage, is one piece. I had to cut the lower half of each wing from the center section to attach the resin wings, **7**. Once the center section was cut away, I attached it to the fuselage, **8**. I packed the space between the center section and cockpit floor with epoxy putty then drilled

holes in each of the wings and inserted a section of brass rod between the two. This allowed me to adjust the dihedral of the wing before I glued it into place. Once the epoxy putty hardened, I anchored the wing assembly to the fuselage. After I was satisfied with the fit, I filled the area with gap-filling super glue followed by a thin layer of putty and sanded it smooth, **9**.

The basic conversion wasn't difficult. I followed my father's old adage of "measure twice, cut once," and I was surprised it went so smoothly. I cut the tail section and replaced the kit's tail and rudder with the broader fin supplied in the conversion, **10**.

The elevators on the Mk.XIV are shaped differently than previous versions, so I cut the elevators from the kit parts and shimmed the edges with styrene stock to ensure a proper fit, **11**.

I attached the tail and rudder and added the horizontal stabilizers. I wanted a relaxed look to the rudder and eleva-

tors, so I glued the rudder into place at an angle and decided to drop the elevators. I left them off until I finished painting, as it would be difficult to paint the tail with them lowered.

Before I applied the primer, I sat back and looked at my project. Amazingly, there wasn't much left to the original Hasegawa kit, **12**.

Seeing things in black-and-white. I marked the camouflage pattern onto the wings and fuselage using the Aero Details book as a reference, **13**. Rather than prime the model with a gray or white primer, I gave the entire model a coat of Future acrylic floor polish before applying the Polly Scale acrylic British dark green and ocean gray, **14**. I've always had good luck with acrylics adhering to Future, so I decided to try something different because I knew this model would be a masking nightmare.

Only a few squadrons were operating

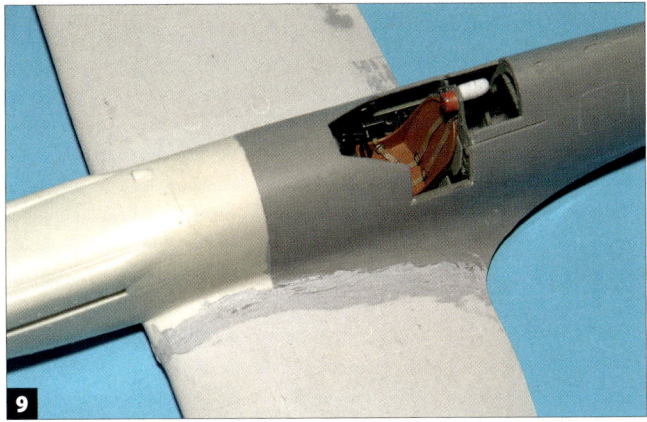

9 After checking the dihedral of the wings, Jeff applied gap-filling super glue to the joint followed by a layer of putty.

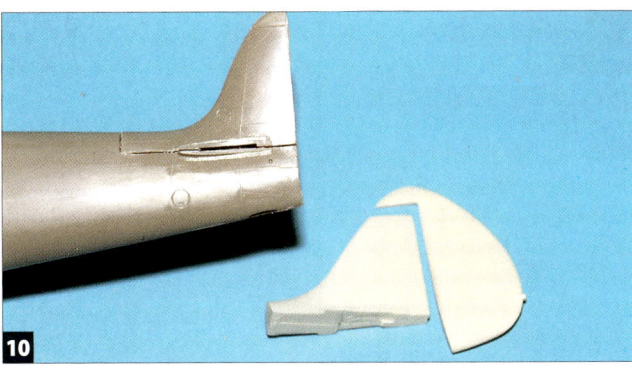

10 Jeff cut the kit's tail with a razor saw and added the larger resin replacement. Griffon-engined Spitfires had larger rudders to compensate for the increased torque caused by the more powerful engine.

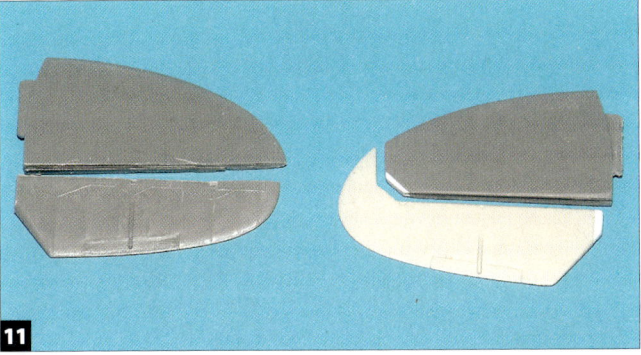

11 Each of the stabilizers was modified to accept the larger elevators.

Right: The center fuselage and wing tips are all that remain of the original Hasegawa kit.

12

Spitfire Mk.XIV's in June 1944, and even fewer were operating over the Continent during Operation Overlord. A few photographic references show 130 Squadron Spitfires wearing D-day stripes, so I was limited in my choices. Undaunted, I masked and sprayed the white stripes on the wings and fuselage, **15**. The blisters for the 20mm wing cannon happened to fall right on a stripe-demarcation line, and I knew that masking a straight line over a compound curve was nearly impossible. Then, I remembered a tip given to me by a friend who mentioned that heating Tamiya masking tape with a hair dryer will increase the flexibility of the tape. With nothing to lose, I warmed the tape with a hair dryer, set it down over the wing blister, and burnished it into place. Amazingly, it worked!

After I finished the white stripes, I sprayed another coat of Future to seal them before I started painting the black stripes. My hunch was correct – the lay-

ers of Future prevented the fragile acrylic paint from lifting, even with burnished down masks, **16**.

After the D-day stripes dried, I masked the leading edges and sprayed the wing ID bands with white primer, **17**. The white primer provided a solid, bright base coat for the yellow and prevented the bleed-through of the green and gray, **18**. After I removed the mask, I had a bright yellow stripe, **19**.

"Sky" writing. I originally intended to use custom-made decals, but unfortunately, something went terribly wrong, and my home-made decals disintegrated. I needed sky-colored code letters A, P, and Y but not much is available in 1/32 scale. And remember, no 1/32 scale Spitfire Mk.XIV kits are available. I rummaged through my loose-leaf binder of decals and masks and happened to find a set of vinyl masks for the Trumpeter P-40. Across the underside of the aircraft is U.S. ARMY, and as

luck would have it, they were the proper height for my codes. I masked the tail of the R to make it a P and cut away the rest of the letters until I had A, the modified P, and the Y. The font is extremely close to the RAF code letters, the dimensions are spot-on, and best of all, it resolved a crisis! I added the fuselage roundel decals to the model to provide a reference point for the codes, then removed the mask from the backing and attached it to the fuselage. I sprayed the fuselage codes with British sky, **20**, and removed the mask. I applied the aircraft serial number RB165, using decals from my spares box. The wing and fuselage decals came from the kit, but I wasn't thrilled with the red – it appeared too bright for me. My choices were, however, limited so I applied them to the model with a little setting solution.

Finishing touches. The remainder of the build was straightforward. I attached the prop, landing gear, wing guns, antenna,

Jeff marked the camouflage pattern onto the wings and fuselage using the Aero Details Spitfire reference.

Jeff sprayed the model with Future floor polish before and after the base coat was applied. This prevents the paint from lifting during masking.

The white stripes were applied to the wings and fuselage, followed by another coat of Future.

After the Future dried, Jeff applied the black stripes and touched up the leading and trailing edges of the wings.

and exhaust stacks, then sprayed the model with Future mixed with Tamiya flat base (X-21) in a 5-to-1 ratio of Future to base. This gave the model a flat, even finish. I attached the canopy with Micro Kristal-Kleer and used monofilament for the radio wire.

This was one of the few projects I hated to see end. It recalled my early modeling days when I was so excited about a project throughout the course of the build that I rushed to get it done just to see the result.

REFERENCES

Aero Detail 30: Vickers Supermarine Griffon Spitfire Dainippon Kaiga Co., Ltd., 2001

Spitfires and Polished Metal – Restoring the Classic Fighter Graham, Modd & Barry McKee, MBI Publishing, 1999

The Supermarine Spitfire Part 2: Griffon-Powered Robert Humphreys, SAM Publications, 2001

British Aviation Colors of World War Two Arms and Armor Press, 1976

Spitfire: RAF Fighter Dan Patterson & Air Vice-Marshal Ron Dick Howell Press, 1997

The Spitfire V Manual RAF Museum Series, Greenhill Books, 2003

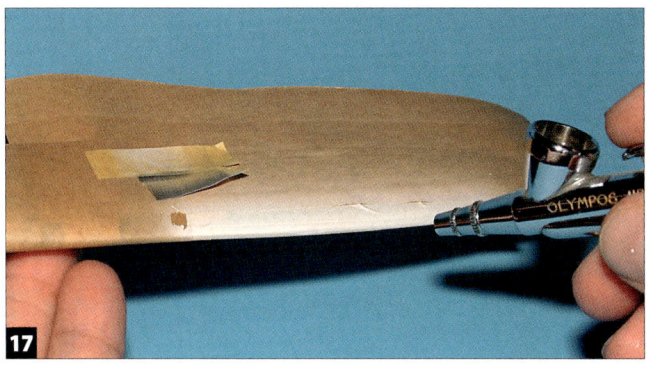

17

A white primer coat on the leading edges prevents the bleed-through of the base colors …

18

… and provides excellent coverage for the yellow leading-edge ID bands.

19

The yellow ID bands and D-Day stripes helped identify Allied aircraft during the Normandy landings, from the ground, above, and dead-ahead.

20

After Jeff's decal experiment crashed and burned, he adapted a set of U.S. Army vinyl masks for the squadron codes.

As well known as the Spitfire is to modeling fans, there's still no 1/32 scale Mk.XIV available, but at least there's a conversion kit!

Painting invasion stripes on a 1/48 A-20

It's not all black and white   BY LARRY SCHRAMM

PHOTOS BY JIM FORBES AND WILLAM ZUBACK

Larry's A-20G features full invasion stripes and decal markings from AeroMaster.

Planning and designing the identification system for the aircraft of the Allied invasion force took months, but applying the stripes was done in less than three days. You can imagine the chaos; the stripes were applied with whatever paint and tools were handy.

Some stripes were carefully masked and sprayed, others slopped on with mops! Most photos don't show how crude some of these stripes were.

When it comes to painting these identification markings, modelers come in two stripes (sorry, I couldn't resist): those who find painting black and white a snap, and those who find alternate markings for their model so they don't have to paint the stripes.

You're probably saying, "Big deal: Spray the white, mask, spray the black, done."

Well, that's basically correct, but there are a few different ways to go about it.

Depending on the subject, I sometimes paint the overall colors first and then add the stripes, or paint the stripes first, then mask over them and apply the overall scheme. Either one is effective, so you can choose which works best for you. I'll show both methods on a 1/48 scale AMT/Ertl A-20G Havoc.

WING METHOD 1

The first thing I do is cut masking tape into strips. The invasion stripes for multi-engine aircraft were 24" wide (18" for single-engine fighters), so for my 1/48 scale A-20, I cut a bunch of ½" strips from blue painter's masking tape. To get sharp edges, I applied the tape to a piece of plate glass and cut with a sharp blade. I also cut long thin strips, about ⅛" wide; we'll discuss them later.

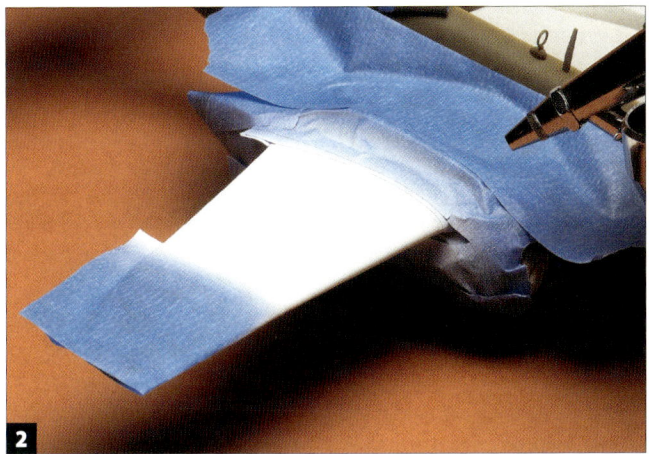

If you're painting the stripes after the overall colors, determine the areas where the stripes should go. On wings, start masking inboard and work out; on fuselages, start forward and work back. Use a thin strip of tape to mask a straight line around a compound curve or irregular surface. The thin strip is more flexible and can "make the turn" around curves.

To ensure the proper width of the area to be painted, lay five short pieces of the precut ½"-wide tape on the model. Mask outside the last strip, then remove the spacers.

Paint the entire stripe area flat white. Caution: If you don't want the white/black contrast to show through an overlying decal, mask the area where the markings go before painting the white. After the white paint is dry, apply five pieces of precut tape (½" in this case) over the white-painted areas. Lay each piece tightly next to the other, with no gaps. Laying all five pieces down will establish proper spacing.

Remove the tape from the areas that will be painted black. Burnish the remaining masking strips to keep the black paint from creeping underneath the edges.

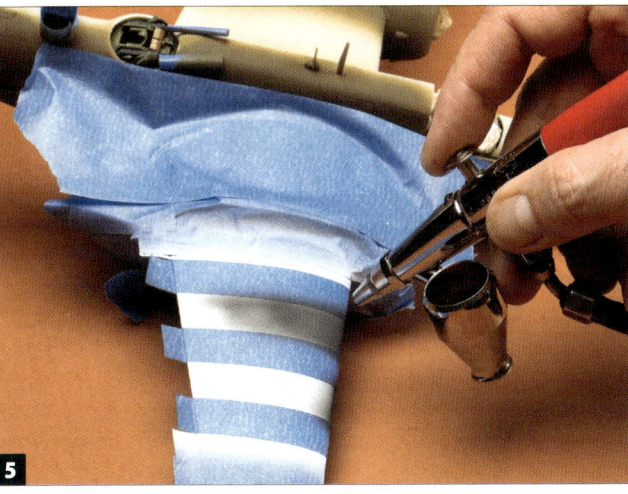

Spray flat black on the exposed areas. Don't flood the paint on; wet paint can creep underneath the masking tape and create a messy and jagged edge on the stripes.

WING METHOD 2

The method I use when painting the stripes before the overall paint scheme is similar. One of the differences is that the white area can be sprayed on without masking. Paint the entire area (or sometimes the entire model) white and let it dry. Follow wing method 1 to mask for the black stripes.

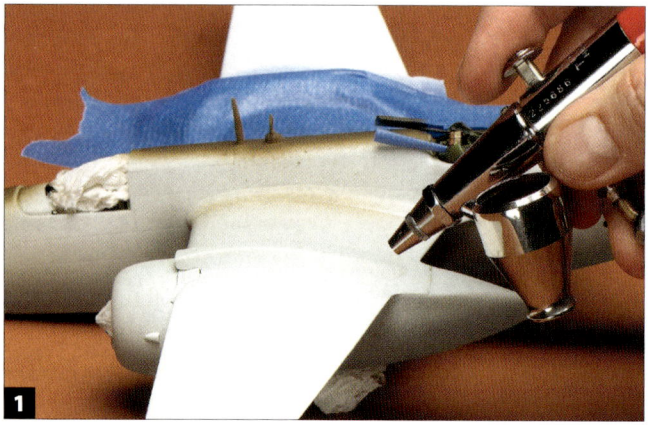

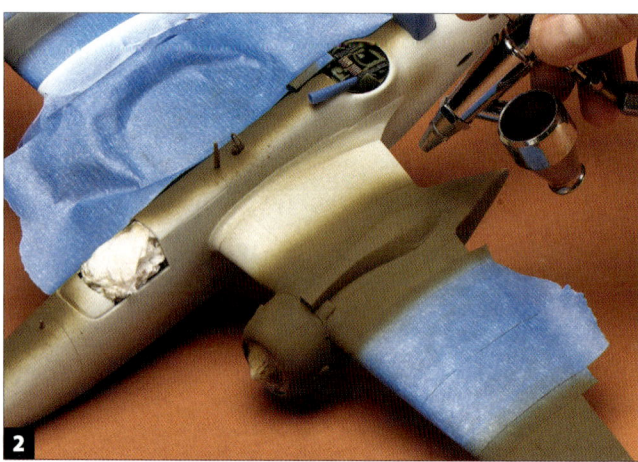

After the black stripes have dried, establish the outside edges of the striped area with thin strips of tape, then mask the rest of the area with big pieces of tape. Spray on the camouflage colors, and when they are dry, remove the masks.

In the case of this A-20, I cut a mask for the location of the national insignia. This can be done by laying low-tack frisket film (a clear self-adhesive masking material for airbrush painting) over the insignia decal on its sheet. Trace over the decal below the frisket with a sharp blade guided by a straightedge and a circle template. Don't press too hard; you need only cut through the frisket material, not the backing sheet.

Peel the frisket from the sheet and apply it to the model in the position where the decal will go. In this case, I used the surrounding part of the frisket to mask over the stripes, then painted the olive-drab camouflage. If you use the first masking method, the frisket in the shape of the insignia would be used.

It's easy to see from this view how visible the invasion stripes were to the Allied troops in Normandy.

FUSELAGE METHOD

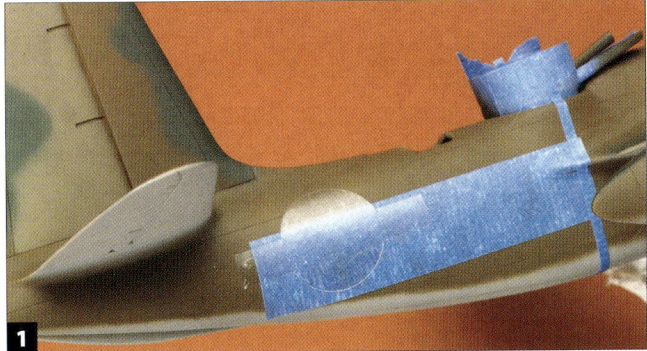

Fuselages can be tricky. Compound curves and wing fillets can make it difficult to get straight circumferential lines. On the A-20, the rounded fuselage is complicated by a tapered cross section. The first thing I masked was the area for the national insignia and the fuselage code letters. Here you can see the frosty clear frisket material and a blue tape strip over the pre-camouflaged fuselage. The thin tape strip under the turret wraps over the wing trailing-edge fillet and around the fuselage, establishing the forward edge of the invasion stripe area.

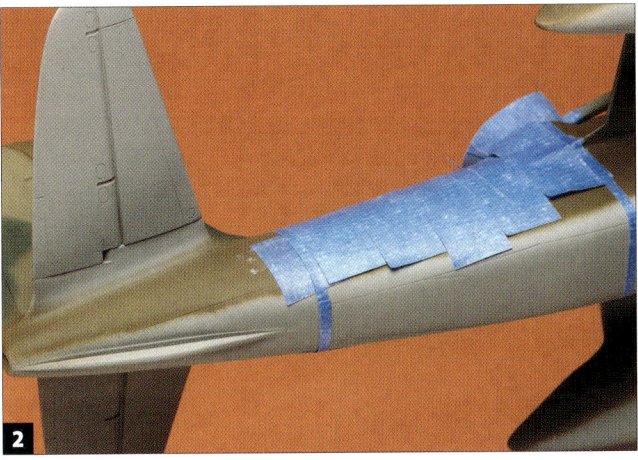

Spacers from ½" tape are positioned to determine the aft end of the stripe area. Another thin piece of tape was worked around the fuselage at that point. Creating a straight line around the fuselage with a wider piece of tape would be nearly impossible. After masking the outside, the spacers were removed and the area was painted flat white.

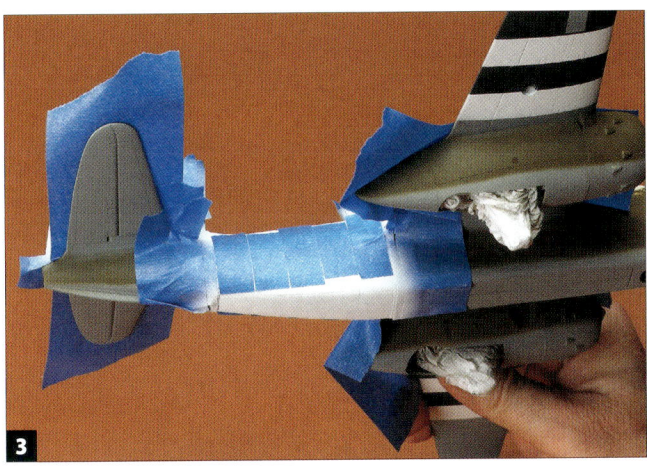

Spacers were once again placed on the dry white paint to establish the pitch of the black stripes.

Thin strips were applied starting at the spacers and wrapping around the fuselage.

Wider strips of tape "backfill" the areas to remain white. It's time to apply the flat black paint.

Here's the finished fuselage. The stripes are vertical even though they go over the tapering rounded fuselage and fin fillet. Note the area for the insignia and code letters were not striped.

41

Build a better BELLE

Combining kits helps build a 1/48 scale museum-quality replica

STORY, PHOTOS, AND ILLUSTRATIONS BY LOUIS ARMOUR

The *Memphis Belle* was the first B-17 in World War II to complete 25 missions with its
original crew. Louis combined parts from B-17F and B-17G kits to build a
museum-quality *Belle.*

*M*emphis Belle – has there ever been a more romantic-sounding name for a B-17 Flying Fortress?

Along with the B-29 *Enola Gay*, it is one of the best-known bombers of World War II. I've wanted to build a model of the *Belle* for years, but building an accurate replica wasn't an out-of-the-box project.

Revell-Monogram makes 1/48 scale kits of the B-17F and the later B-17G. The B-17G kit (No. 85-5600) has superior details and a complete interior, but the B-17F kit (No. 85-4701) comes with *Memphis Belle* markings and has the correct window arrangement and nose cone. Combining both kits was the first step toward building a better model of the *Belle*.

Too much research? No way! A few years ago I traveled to Memphis, Tenn., and photographed the full-size *Memphis Belle* from every angle. After learning that it was built before Boeing had settled on a standard interior arrangement, I was graciously granted permission by the *Memphis Belle* Memorial Association to photograph the interior of the plane on a later trip. The photos and notes I gathered helped me create a much more accurate model. (Since then the *Memphis Belle* Web site has posted a number of interior detail shots; see them at www.memphisbelle.com).

Most of the interior detail is hard to see in the finished kit, so you may opt to just concentrate on the exterior details. If you do decide to detail the interior, spend most of your time on the radio room, as it's the most easily viewed part of the interior. I don't normally spend building time on things I know won't be seen, but while I waited to find answers to my research questions, I just kept fiddling with the interior!

The *Belle* was an early F model and has a different interior arrangement than that included in both kits. Many of the details are the opposite of what they should be. I started by adding stretched-sprue interior ribs to the F fuselage using the G kit as a guide. I painted the finished interior walls Testor Model Master interior green (No. 1715) and gave them a wash of thinned black paint.

I made new side windows for the nose from a clear plastic package that had the right curvature and installed them

with white glue. I used Evergreen plastic strip stock for the gun mounts in the side windows.

After enlarging the gun opening in the windows and nose cone with a twist-drill bit, I installed a small black craft bead to represent the swiveling socket the guns poked through.

I used the interior from the G kit after cutting it in half behind the "step" in the nose compartment. This made things easier to work on until I was finished.

I removed the molded-in chair bases, then plugged and filled the holes. I also plugged and filled the bearing hole for the nose turret and removed its control

(part No. 47). I removed all the molded-in details from the bulkhead (17) and added canvas covers made from two-part epoxy putty. Remember: Most of this work is invisible once the fuselage is sealed up. Have fun, but don't knock yourself out!

The kit's nose guns simply poke through holes in the nose cone. I wanted to accurately model the gun mounts that supported the weight of the .50-caliber machine guns. To do so, I first cut a .010" sheet-plastic ring to fit between the fuselage and the clear nose. After gluing it to the clear nose, I made tiny holes the size of the bracing in this ring with a pin vise and a small drill. To model the braces, I

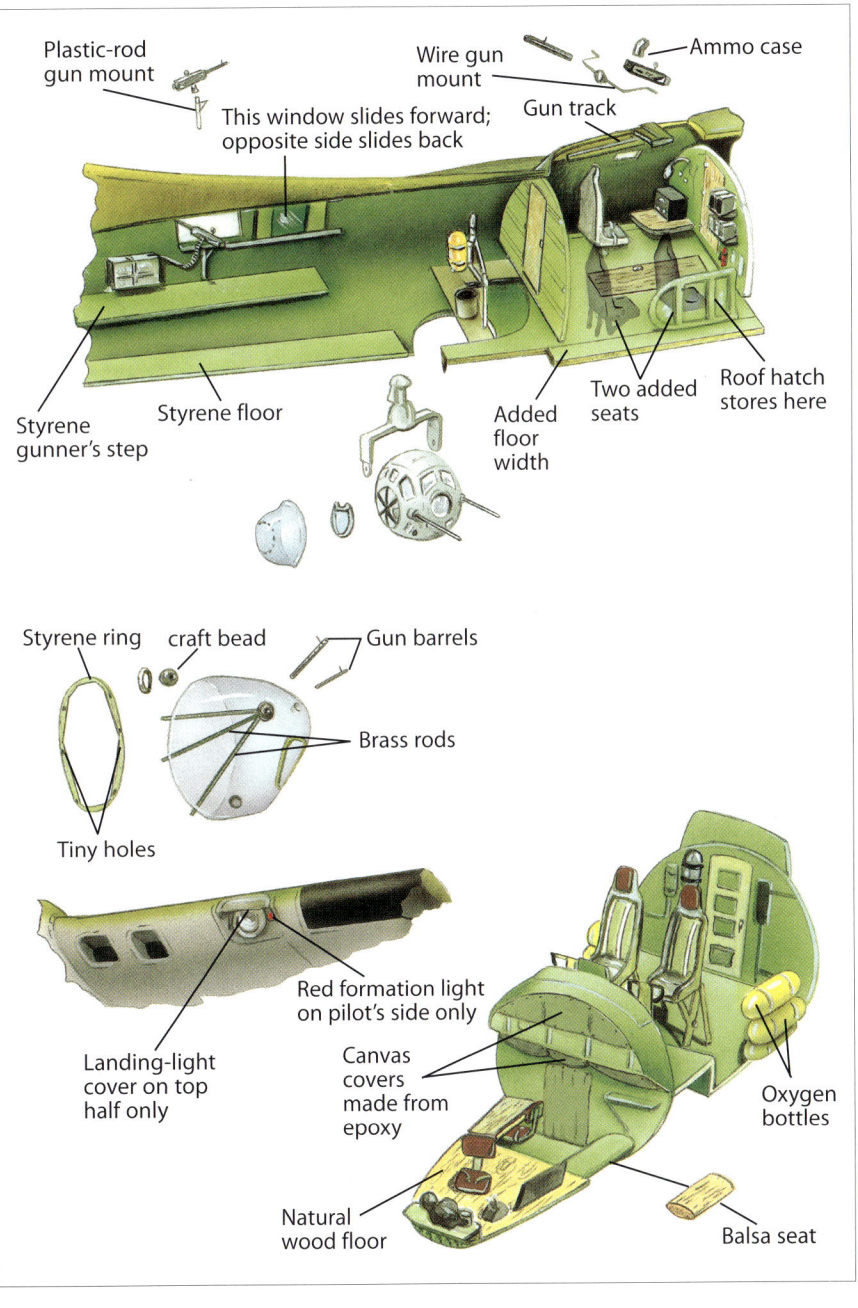

Plastic-rod gun mount

Wire gun mount

Ammo case

This window slides forward; opposite side slides back

Gun track

Styrene gunner's step

Styrene floor

Added floor width

Two added seats

Roof hatch stores here

Styrene ring

craft bead

Gun barrels

Brass rods

Tiny holes

Landing-light cover on top half only

Red formation light on pilot's side only

Canvas covers made from epoxy

Natural wood floor

Oxygen bottles

Balsa seat

painted fine brass rods olive drab, pushed them through the holes, and glued them to the gun sockets.

Then I cut off the excess brass wire flush with the ring. After painting each gun, I cut off the barrel and glued the body to the socket from the inside. I would add the barrels during final assembly, lining them up with the gun bodies inside the nose.

The nose guns had a silver ammo box attached to one side; I used parts from Verlinden's WWII aircraft gun set (No. 1267). When I was satisfied with the interior, I glued the fuselage together and sealed any openings with tape to keep out dust.

I had to modify the wing and horizontal-stabilizer mounts to accept the G version's parts. I was being nit-picky about

details here; you could use the F version's wings and stabilizers without losing much detail. Before gluing the wings together, I opened up the various intakes and exhaust openings with a small twist drill and files. Parts from Eduard's photoetched grilles and details set (No. 48208) were added to the leading-edge intakes. I made the waist-gun mounts from stretched sprue.

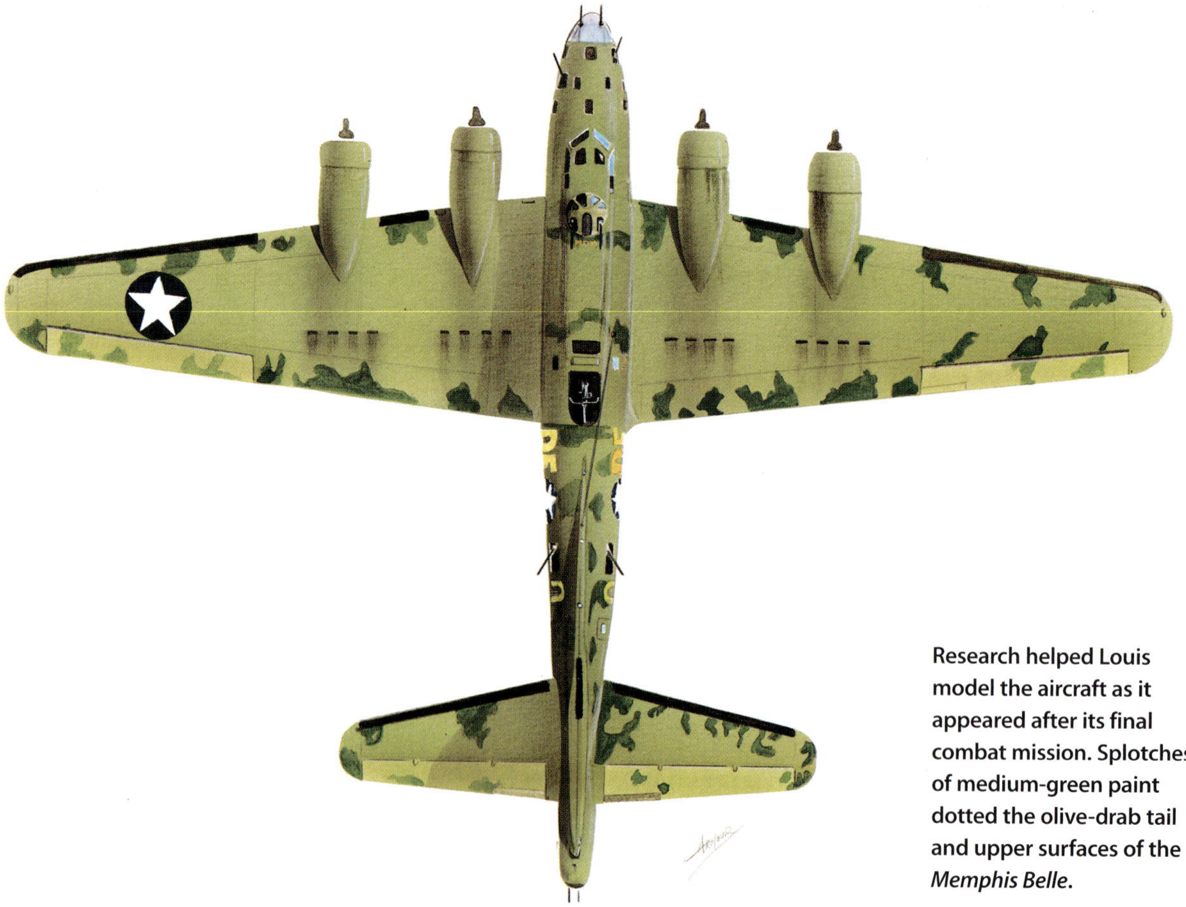

Research helped Louis model the aircraft as it appeared after its final combat mission. Splotches of medium-green paint dotted the olive-drab tail and upper surfaces of the *Memphis Belle*.

I wanted to lower the elevator slightly for a more natural look, **1**. Using a hobby knife with a No. 11 blade, I scored the hinge line repeatedly until I cut through. Afterward, I glued the parts together and filled the opening with sheet styrene. I sanded the front of the separated elevator to an airfoil shape and notched it where the hinges would go. Small styrene chips formed the hinges. In the cockpit, I pushed the control column slightly forward to reflect the elevator's new position.

The radio compartment had a gap between its walls and the floor, so I added a small strip of styrene to each side of the floor. The roof hatch was painted and then placed along the wall opposite the radio operator's table. This is where the roof hatch was stowed when not in place. I added the two pilot seats from the G kit to the right side; they would be used by the gunners during takeoff. In the kit, the radio operator's table is a large block of plastic. I cut it off, filled the opening with sheet plastic, and made a new table from .040" styrene sheet. I added a small goose-neck lamp over the radio table.

I littered the floor with shell casings made from small pieces of copper-painted stretched sprue. I mounted a .50-caliber machine gun in the *Belle*'s radio-compartment hatch opening. I made the mount from stiff wire bent to shape and glued to a small wire ring. I made the small triangular antenna mount on the plane's right side from a scrap of plastic sheet.

Paint and markings. Photos of the *Belle* showed that her markings changed during her career. Even the experts are at odds as to what markings were on the plane at any given time. I wanted to portray the *Belle* shortly after its final combat mission. Some of the markings would be educated guesses.

I started by masking the windows and turret openings with drafting tape. To mask the engines, I used thin foam packing material cut into a circle and pushed into the engine cowling, **2**. After priming the model dark gray to check for surface flaws, I airbrushed Testor Model Master neutral gray (No. 1725) over the underside.

Models of large aircraft painted with one or two basic colors look like what they are: models. Full-size planes are painted with spray guns (sometimes in the field) and are not uniformly colored. Sun-bleached paint, wear and tear, and field-repair work yield paint jobs with a variety of shades.

To represent this, I masked off the gray underside and airbrushed Model Master olive drab (No. 1711) over the upper sur-

The drooping elevators give the *Belle* a natural, "relaxed" look.

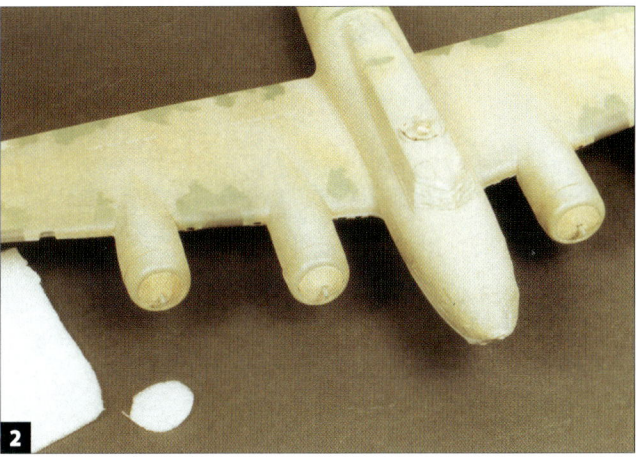

Discs of foam packing material stuffed into the engine cowlings mask the radial engines.

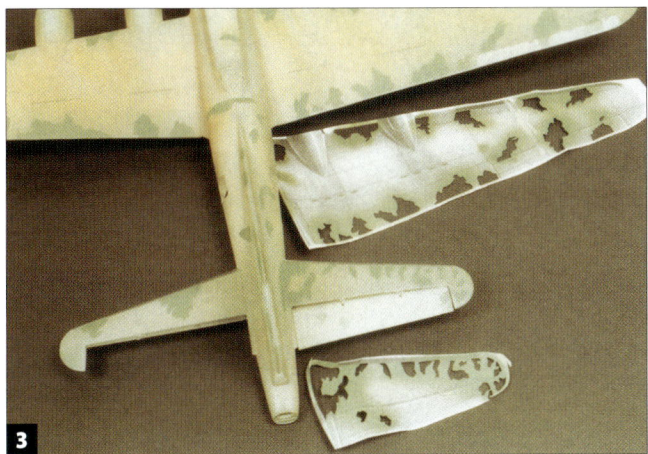

Camouflage diagrams, enlarged on a photocopier, helped Louis apply the *Belle*'s camouflage pattern appropriately.

The model, with its camouflage pattern, during decal application. Louis gave it an overall coat of Future to prepare its surface for decals.

faces. I followed this with another coat, lightened about 20 percent with white paint. I applied this paint to the surfaces of the upper side that would have seen a moderate amount of sun and faded somewhat. Lastly, I sprayed the uppermost surfaces with Model Master faded olive drab (No. 2051). I rubbed some parts of the wings with a plastic scouring pad to wear down the colors, then repainted with slight variations of olive drab.

Rubbing the underside of the model with the pad wore away the paint on the raised details, revealing the darker primer underneath on the details. I wasn't too subtle, as the weathering and Dullcote I would apply later would tone things down and even out the finish somewhat.

Splotches of medium-green paint dotted the tail and upper surfaces of the *Memphis Belle*. After consulting many photos, I decided that the pattern suggested in the F-model instructions was accurate. Using a photocopier, I enlarged the paint-ing instructions to the size of the model and cut out the patterns with a hobby knife. Then I airbrushed Testor Model Master medium green (No. 1713) through the patterns after applying them to the model with double-sided adhesive tape, **3**.

I painted some of the splotches by hand, then carefully softened their edges with the airbrush. I airbrushed the fabric-covered control surfaces with a much lighter shade, as these areas faded faster than the metal parts. Finally, I applied a thin wash of burnt-sienna oil paint to the details on the upper surfaces and allowed everything to dry.

I airbrushed the entire model with a coat of Future to prepare it for decaling, **4**. I used Super Scale's decal sheet (No. 48-59) for the nose art and markings, and used the F-kit's decals for everything else. Both decal sheets err in the same respect – the wheel cover on the pilot-side landing gear was blue, not red. I scanned the decal sheet, and using my computer I changed the red to blue. I printed a new wheel cover decal on blank white decal paper using my color copier.

After decaling, I airbrushed a light coat of Floquil dust (No. B-110006) over the decals to give them a faded look. Then I applied Testor Dullcote (No. 1106) to the entire model to give it an even, flat finish that would be ready for weathering.

Weathering and final details. For me, artist's pastels are the most easily controlled and subtle form of weathering available. Using photos of the full-size plane as a guide, I applied a mixture of gray and black pastel dust to the model's panel lines, then removed all but a trace to accent the detail. Black pastel was used to represent soot from the engine exhausts, **5**. I did not seal the pastels with a clear coat since I would not be handling the model once it was attached to the base. The pastels are rather durable, and with careful handling, not prone to wear off.

5

Louis weathered the model with artist's pastels. This photo shows the difference between the factory-fresh left wing and the realistically weathered right wing.

6

Monogram and Verlinden figures were combined and re-cast to model the *Belle*'s crew.

7

The re-posed crew figures reenact a scene from the 1944 William Wyler documentary, *The Memphis Belle: A Story of a Flying Fortress.*

8

A 1/48 scale Verlinden "follow me" jeep rounds out the scene and helps give it a sense of scale.

I used a clear craft jewel for the wing landing light. I made a cover for the light from a clear box cover and attached it with white glue. Note that the light cover on the *Belle* covers only the top half of the light. I used small 1.5mm-diameter jewels for the formation lights and 2.3mm-diameter jewels for the red and white lights on the tail. The final touch was to add the antennas and gunsights, which I made from stretched sprue.

The base and figures. I used a piece of ¾" plywood slightly larger than the plane for the base. I applied dark-walnut wood stain to the edges and applied a coat of polyurethane to seal everything. I cut a piece of pebble-textured mat board to the size of the plywood and attached it with white glue. Heavy books placed on top kept everything flat until the glue dried. I painted the surface brownish tan. Marking off the expansion joints with a ruler and a pencil, I applied black acrylic

paint to the lines with a hypodermic syringe.

I cut a mask with a square opening the size of one of the base's concrete slabs, then airbrushed dark gray paint through it onto the edges of each square. I darkened some squares completely to add some variety to the surface. Tire marks were applied using pastels and a brush the width of the tires. Using thinned black paint, I added oil spots and drips under the two outboard engines, as my reference photos showed large oil stains under them.

The 14 crewmen were made from Monogram and Verlinden figures. I made a silicone mold of four basic figures, **6**. After casting new figures in resin, I cut them apart and posed them as needed. I turned heads, moved arms, and added stretched-sprue cigarettes. One guy even has a box camera made from plastic scrap. I avoided painting any two items of clothing the same shade, as "real" clothing wears and

discolors differently. The figures give the finished model a sense of scale.

I posed the men in a scene with a crew member being hoisted up to kiss the "Memphis Belle" painted on the plane's nose – this real-life event can be seen in the *Memphis Belle* documentary, **7**. A 1/48 scale Verlinden "follow me" jeep rounds out the scene and helps give it a sense of scale, **8**.

In one corner of the base, I attached an enameled-brass plaque that the surviving crew members signed with a gold paint pen. I also attached a small nameplate identifying the model.

I'm pleased to say that at the King Con model show in Memphis, Tenn., the president of the *Memphis Belle* Memorial Association carefully examined my work. He then asked me if I would donate the model to be displayed with the full-size plane in her new, soon-to-be-built, permanent home. That privilege made all the effort worthwhile.

Louis' first-hand research on the full-size *Memphis Belle* helped him create a much more accurate model inside and out.

Mosquito
AMERICAN STYLE

A green modeler turns a 1/48 scale Mosquito PRU Blue   BY STEVE RICHARDS

Kitbashing Airfix and Monogram Mosquitos with Paragon resin two-stage engines makes this colorful USAAF Mosquito P.R. Mk.XVI.

A few years ago, watching a short clip of black and white World War II footage inspired me to build a model of an 8th Air Force Mosquito. After research, I found photos – some in color – of the Mosquito P.R. Mk.XVIs in service with the 25th Bomb Group (Recon).

No straight-from-the-box, 1/48 scale Mosquito kit will yield this version, so I decided to make one of these colorful "Mossies" for my first conversion. Paragon had resin engines to fit the Airfix kit, but the Monogram kit had the glass nose necessary for this version. I decided to use the Monogram fuselage and tail and the Airfix wings. (If I were to do this conversion today, I would use the Paragon engines and accessories on the recent Tamiya B. Mk. IV kit instead of kitbashing.)

Here is my shopping list:

- Paragon two-stage Merlin engines (resin set No. 4842)

- Paragon 100-gallon external wing fuel tanks (No. 4878)

- Paragon PR-type canopy (No. 4873)

- AeroMaster decal sheet (No. 48-082)

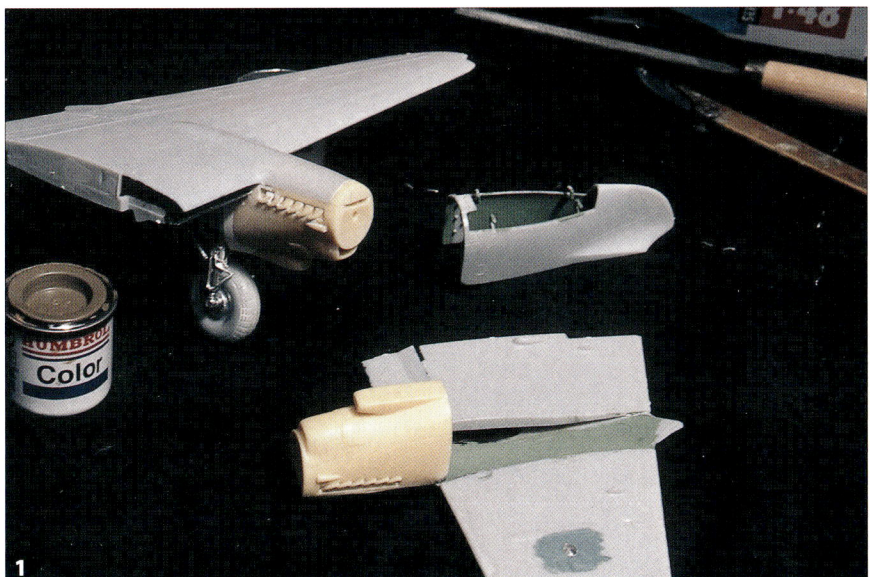

1

The Paragon engines, designed to fit the Airfix wings, are mounted to the top surface. The remaining Airfix nacelle (right) awaits attachment.

2

Sanding the epoxy-putty-filled seams on the assembled nacelle is next.

Engine change. The major difference between the kit Mossies and the P.R. Mk.XVI was the shape of the Rolls-Royce Merlin 72 two-stage engines. The Merlin 72 nacelles were slightly longer and shaped differently. The Paragon engine nacelles capture this shape, but you have to remove much of the Airfix nacelles to fit the new ones.

First, I glued the Airfix engine nacelle halves (part Nos. 51, 55, 59, and 63) and allowed them to dry. I positioned masking tape as a cutting guide 33mm from the front of the engine. This cut leaves the forward section of the wheel well intact, but a bit fragile!

I carefully sawed down the line starting at the underside of the nacelle. A razor saw is ideal here. Next, I cleaned the rough-cut edges with sandpaper.

My plan was to glue the resin engines to the top surface of the cowl, which is part of the top wing half, **1**. Then the assembled rear portion of the nacelle (with the wheel well) is added underneath the wing. First, I had to remove the resin pour stub at the rear of each engine. The engines are marked R (right) and L (left); don't mix them up.

Plastic cements have no effect on resin parts, so you must use super glue, epoxy, or epoxy putty to bond resin to plastic. I used super glue, then filled the seams with Milliput, a two-part, fine-grained epoxy putty. It sets hard, but with progressively finer grits of sandpaper it sands easily to

3

A coping saw was used to remove the wing mounts from the Monogram fuselage halves before attaching the Airfix wings.

a velvet-smooth finish. Plastic-to-plastic joints were made with liquid cement.

Adding the landing gear was tricky, as it was designed to be trapped between the nacelle halves as you assemble them. Since I had already assembled the nacelles, I had to modify the mounting system so I could anchor the struts. I figured this would be better than trying to cut the nacelles with the gear in place!

As the two-stage Merlin was longer than the kit's engine, Paragon provides a resin half-moon extension for the top of the cowling. This was added last, **2**. Paragon also provides photoetched grilles for

the intakes, and these were fastened with super glue.

Fuselage surgery. At this stage, my morale was riding high; the hardest part of the conversion had gone better than I expected. However, it was immediately clear that the Airfix wings could not mate to the Monogram fuselage because of different assembly methods. I was faced with cutting out Monogram's wing mounts with a coping saw, **3**. This tool allows detaching the blade from the bow so you can run it through a pilot hole and reattach it for cutting.

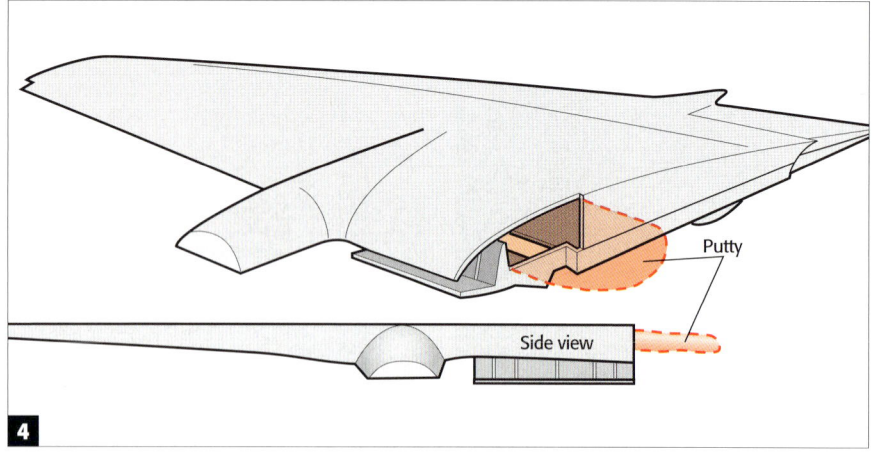

4

An epoxy-putty plug in the leading edge of the Airfix wing establishes the position of the leading edge in the cut-out Monogram fuselage.

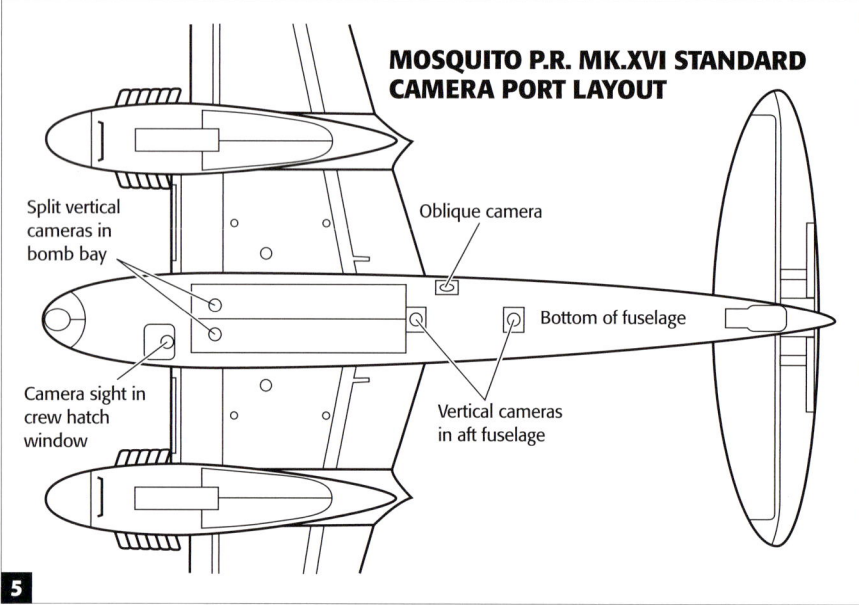

MOSQUITO P.R. MK.XVI STANDARD CAMERA PORT LAYOUT

Split vertical cameras in bomb bay

Oblique camera

Camera sight in crew hatch window

Bottom of fuselage

Vertical cameras in aft fuselage

5

The underside of a standard P.R. Mk.XVI had five camera ports installed, but it is unclear from photos whether MM 345 had these ports.

Since the wing/fuselage assembly arrangement was changing, I had to modify the Monogram cockpit floor as well. The bomber cockpit floor includes the bomb bay, and since I was going to have the bay closed, I cut this portion away (up to the radio, sited behind the crew) so it wouldn't encroach on the modified wing/fuselage assembly. I painted the interior British Interior Green and affixed in the floor and side cabin windows.

Wings to fuselage. Because the wing/fuselage assembly was such a challenge, I didn't close the fuselage just yet. The method I used seems complicated, but it's the best way for this situation. First, I made a location plug from a blob of Milliput and inserted it into the opening in the Airfix wing's leading edge, **4**. When it's set, this plug establishes the position of the wing in the cutout of the Monogram fuselage.

Now I could position the wing with the plug against the front of the cutout. I rolled a long "worm" of Milliput and pushed it from inside the fuselage half around the Airfix wing root. Excess putty that oozes outside the fuselage fills the gaps there, and can be easily cut away before the putty sets.

Before the putty set, I placed the fuselage halves together and secured them with rubber bands – no glue yet. Then I adjusted the wings to the correct angles and left them to cure for 24 hours. When I came back to the project, the wings were set solid and in perfect position.

Smile for the cameras. Cementing the fuselage halves and tailplane assembly was straightforward, leaving me free to drill out the five camera ports. My references show two ports along the underside center line ahead of the tail wheel, one offset to the left side of the rear fuselage, and two more on either side of the forward end of the bomb bay, **5**. I first drilled pilot holes with a 1.5mm bit, enlarged them with a 3mm bit, then cleaned the holes with a needle file. After the model was painted, I filled the holes with Microscale Kristal Klear.

Painting and markings. The model was painted overall Humbrol PRU Blue (No. 124). To produce the slightly orangy red of the tail group, I painted it with satin red (No. 132). Only the underside of the fuselage carried D-Day stripes, so I painted the area white and used solid-color decal for the black stripes.

When these aircraft were transferred from RAF stocks, the British roundels were freshly overpainted. To produce this effect, I added a little black to PRU Blue, then overpainted the Airfix roundels and applied them to the model. I painted the propeller spinners dark blue (No. 134).

The AeroMaster decals went on after a coat of clear satin to smooth out the finish. After the decals were dry, I overcoated with clear flat on the PRU Blue and another coat of clear satin on the red tail. At this point, I added the slipper tanks to the wings as they were added over the insignias on the real aircraft.

The exhaust streaks along the cowl sides were added by dry-brushing black and dark gray. Dip an old brush into the paint, then rub the pigment off onto paper or a rag. Keep rubbing until the pigment seems to be dry, then repeatedly (and gently) stroke the brush on the model. The remaining pigment will gradually create an indistinct smear. Repeat the process until you obtain the effect you want.

Battle of the bulge. The canopy of the P.R. Mk.XVI had an additional blister on top for the navigator to determine a star "fix," but this is not included in the Monogram clear part. I had obtained a Paragon

USAAF MOSQUITOS

The 25th Bomb Group (Recon) was activated in early August 1944 and assumed the assets of the 802nd Recon Group (Special) (Provisional), a unit that had been pioneering special missions for the Eighth Air Force in England since March 1944. The unit began life as the 8th Recon Group (Special) (Provisional), but was quickly redesignated to avoid confusion with a similarly numbered unit then serving in India.

While the 25th flew a variety of aircraft, it is best known for its two squadrons of Mosquitos. The choice of de Havilland's marvelous wooden aircraft was simple: no American aircraft offered the same range and speed.

During 1943, the Army Air Forces managed to secure 40 unarmed Mosquito bombers from Canadian production; designated "F-8" by the AAF, the aircraft suffered a variety of problems. Only one saw combat (in Italy with the 15th Photo Squadron, 3rd Photo Group), 26 were returned to the RAF, and the rest were wrecked or salvaged.

The British-built Mosquito P.R. Mk.XVI was far more successful, offering the Eighth Air Force the performance needed for its special reconnaissance missions.

The first of 142 of these aircraft delivered to the AAF arrived in February 1944. With the 25th, the Mozzy (as the Americans spelled the British nickname "Mossie") flew a variety of missions, including night flash photography, radar

The subject of Steve's conversion, this colorful USAAF P.R. Mk.XVI rests under scattered low clouds at Watton, Norfolk, England. Photo by Robert Astrella via the Jeff Ethell Collection

mapping, motion picture documentation, LORAN electronic navigation calibration, special OSS communications, target weather recon, electronic countermeasures, and bomber formation command.

MM345, the subject of Steve Richards' model, was delivered on March 2, serving as a transition trainer until her first operational weather reconnaissance mission on Aug. 12. She flew 30 operational missions before a crash landing on Dec. 7, 1944; after repairs, she was again used for transition training until a second crash, on Feb. 20, 1945, ended her career with the AAF.

As with other P.R. Mk.XVIs flown by the AAF, MM345 arrived with RAF markings and overall PRU (Photo Reconnaissance Unit) Blue camouflage. The RAF roundels would quickly be overpainted, with American national insignia added; the RAF fin

flashes were also overpainted, replaced with a white ring (for the light weather squadron) and white individual aircraft letter "Z." Full invasion stripes were applied in early June 1944, then overpainted on upper surfaces later in the month. In August, just before the first operational mission, the group's new crimson tail markings were added, overpainting the white ring, but leaving a PRU Blue disc around the white "Z."

This color photo was taken shortly after the Mozzy's 14th mission, a weather reconnaissance flight over the North Sea on Sept. 14, 1944. By that time, the invasion stripes had also been removed from the underside of the wings. The underfuselage stripes may have been overpainted before the first crash in December 1944; they were certainly gone by 1945.

– Dana Bell

---

vacuum-formed canopy for this project, but it was designed to fit the Airfix kit and wouldn't sit right on the Monogram fuselage. I thought about carving out the top blister and adding it to the Monogram canopy, but instead I cut one down from the dome of Matchbox's 1/72 scale Beaufighter. It was attached to the main canopy with white glue.

I'm proud of my first conversion, and I think the photos show that it looks right. If you've never picked up a razor saw or attempted to use resin parts, I hope that I've encouraged you to move into new pastures!

REFERENCES

Mosquito in Action (Parts 1 and 2) Jerry Scutts, Squadron/Signal Publications, Carrollton, Texas, 1992, 1993

WWII War Eagles, Global Air War in Original Color Warren M. Bodie and Jeffrey Ethell, Widewing Publications, Hiawassee, Georgia, 1995

The de Havilland Mosquito, a Comprehensive Guide for the Modeller Richard Franks, SAM Publications, Bedford, England, 1998

Modeling a MARAUDER

PROJECT 10

Easy techniques turn Monogram's 1/48 scale diamond in the rough into a real gem BY RAFE MORRISSEY

PHOTOS BY RAFE MORRISSEY AND FLOYD WERNER JR.

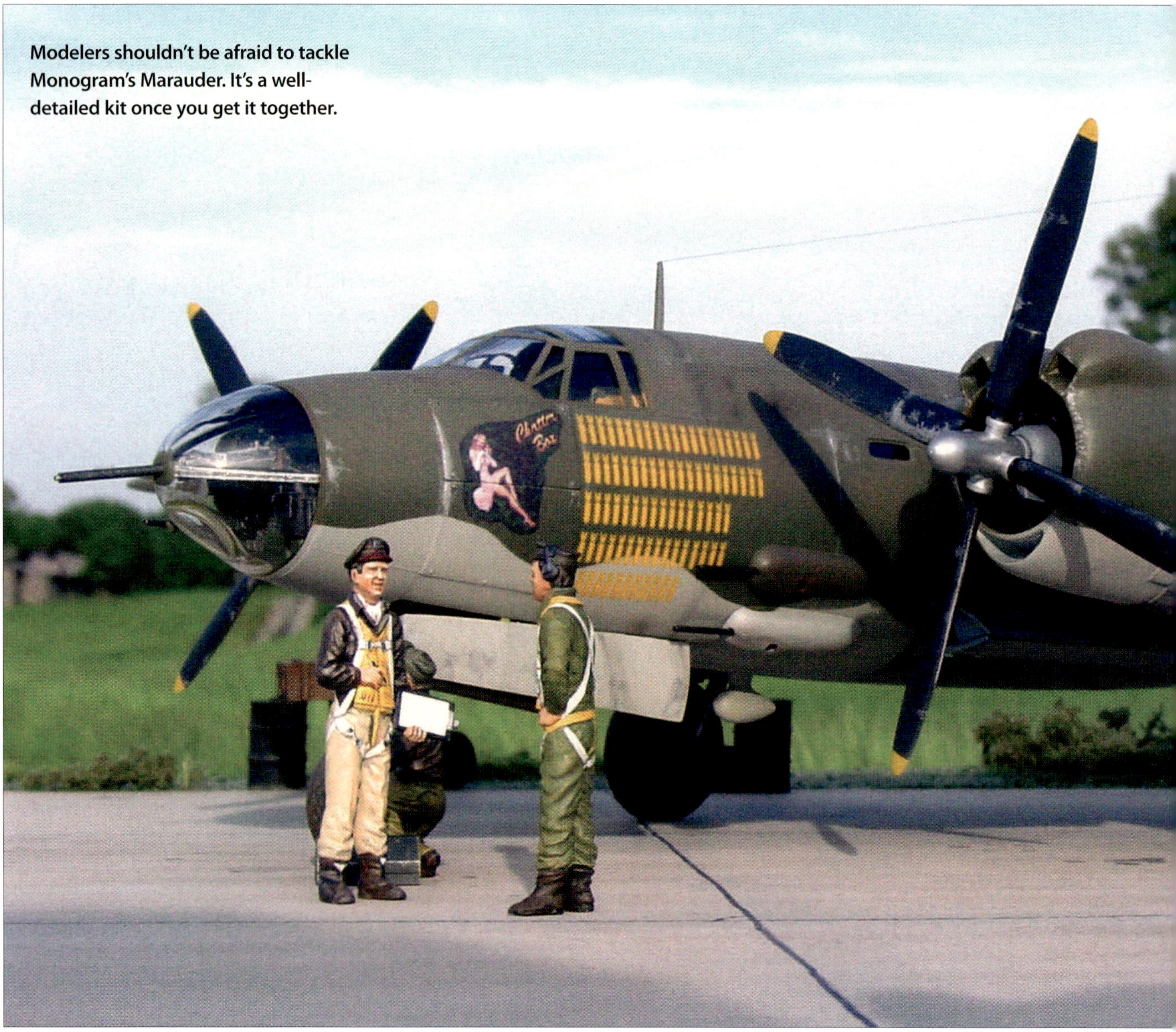

Modelers shouldn't be afraid to tackle Monogram's Marauder. It's a well-detailed kit once you get it together.

Just as pilots were at first reluctant to fly the B-26 Marauder, so too have many modelers been hesitant to build the 1/48 scale Monogram kit. Even though it is eagerly sought by collectors and usually commands top dollar at swap meets and on eBay, assembled models are rarely seen on contest tables or in modeling magazines.

Monogram kits of the 1970s have earned the reputation of being well-detailed but ill-fitting – and the B-26 qualifies in both areas. However, most of the kit's problems are easily fixed, and with some skill and perseverance, you can make it into a model worthy of inclusion in any World War II collection.

The good, the bad, and the ugly. The first step in building the B-26 (or any airplane model) should be to remove the major airframe components from the sprues and tape them together, **1**. This dry-fitting helps identify warped parts and areas where filler will be needed. To fix a warped part, place it under a stream of hot tap water while bending it against the warp. Heating the part allows you to bend it without breaking it. Holding the part slightly past straight while it cools will usually compensate for the plastic's "memory" of its warped condition.

My kit was warp-free, but still had plenty of fit "issues." I made a list of everything that needed to be fixed and noted them on the instructions so nothing would be forgotten later.

Accentuate the positive. One of the best things about Monogram kits from this period is the interior detail. The proper combination of painting and weathering techniques can make this kit sparkle. The first step is to lay down a good base coat. I airbrushed Testor Acryl interior green (FS 34151) for the cockpit and fuselage interior. My research indicated that the bomb bay was painted aluminum or was unpainted. I sprayed a 50/50 mixture of Polly Scale flat aluminum and clear gloss in this area, **2**.

I like to use pastel chalks to apply a "dry wash" to darken recesses – it's quicker than a paint wash. Equal piles of burnt umber and black dust were scraped from the chalk sticks with a knife and mixed together with an old paintbrush. I use an

1
Taping the major pieces together reveals potential fit problems that will have to be corrected.

2
The right fuselage half cabin and bomb bay interiors have been painted.

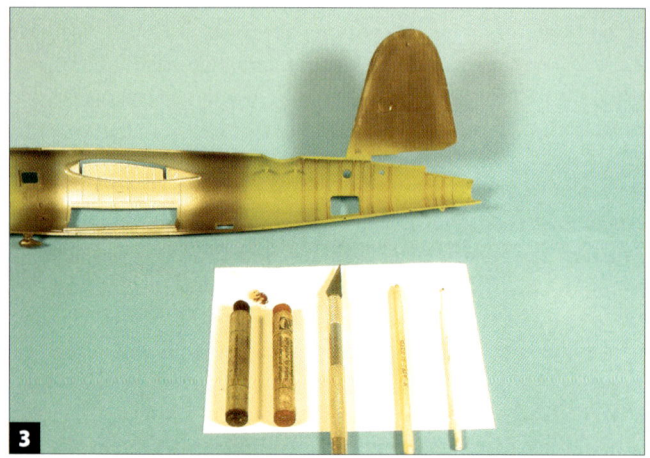

3 Rafe used pastel chalk dust to do a "dry wash" of interior details.

4 The right fuselage half looks much better with the dry-washing and dry-brushing complete.

5 Dry-brushing followed by careful detail painting and a tiny drop of Future makes the instruments look right.

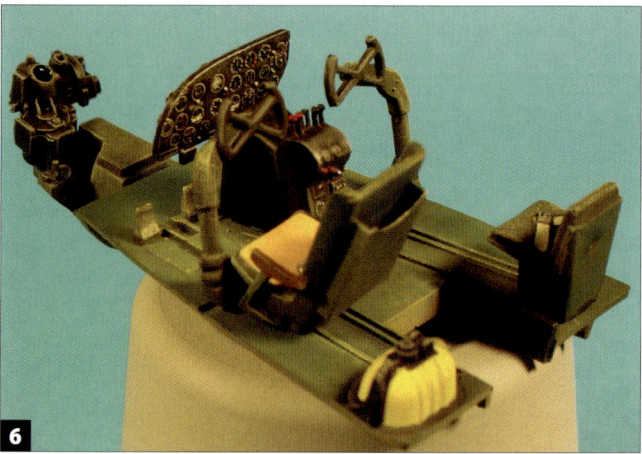

6 Here the cockpit components are tacked together with Blue Tac putty for a trial fit. Adjustments can be made before glue is applied.

artist's blending stump or the smallest Microbrush and run it along ribs, inside recesses, and around raised detail. Making sure to blow away loose dust, I blended the remaining dust into the surface with a cotton swab. The dry wash helps create a shadow effect but allows more control than traditional paint washes, **3**.

Since the dry wash doesn't stick to smooth surfaces, I used diluted raw umber oil paint on the metalized bomb bay. I cut the paint with Testor paint thinner. When it was dry, I wiped the excess away with a cotton swab.

Lightened tones of the base colors were used to highlight raised details. For the interior green areas, I used a 50/50 mix of Testor Acryl insignia yellow (FS 33538) and flat white. I dipped an old ¼" flat brush in the mixture and scrubbed the brush across a paper towel and then a piece of cardboard to remove most of the paint. A few

light passes over the raised detail should transfer a hint of the lightened paint. If you see streaks, you have too much pigment on the brush.

Next, I used a cotton swab to rub dark green pastel chalk dust into the open spaces between the ribs to vary the tone and add visual interest, **4**.

Details, details. For detail painting, I like the new Vallejo acrylics. They're opaque, easily thinned with water, and come in bottles with built-in dispenser tips that allow just enough to be squeezed out. They don't stick well to bare plastic but went on fine over the interior green primer.

I mixed colors for the interior details, but there is a wide array of colors available. I painted each item, then accented shadows with darkened tones and dry-brushed with lightened variations of the

base colors. Finally, I blended everything with thinned coats of the base colors.

I sprayed the instrument panel with a base coat of Testor Acryl flat black lightened a tad with white. I try to avoid using black straight from the bottle as it appears too harsh. Once the base coat dried, I dry-brushed the detail with light gray followed by white. This established the outlines of the instrument faces. I next used yellow, red, and white Prismacolor artist's pencils to tick individual spots on the dials. A dab of Future floor polish over each instrument finished the faces, and a silver Prismacolor pencil helped simulate worn edges on the center console, **5**.

Rather than use aftermarket parts on this project, I wanted to rely on the detail given in the kit. I did make a couple of concessions, though. I wanted the crew boarding hatch open, so I cut it out of the fuselage and made a new folded door

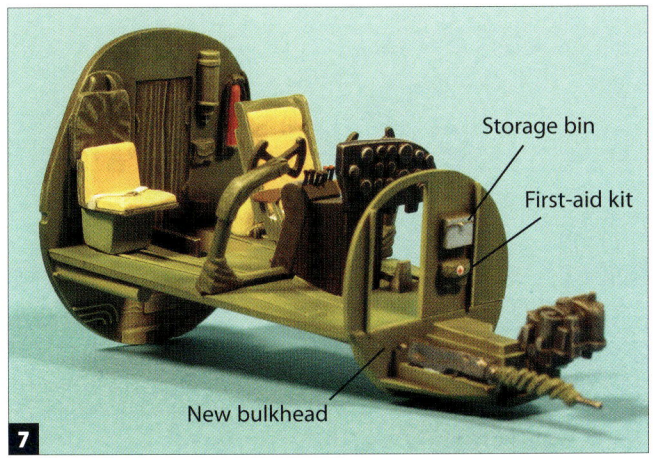

7

Rafe used a contour gauge to determine the shape of the forward bulkhead he added to the interior.

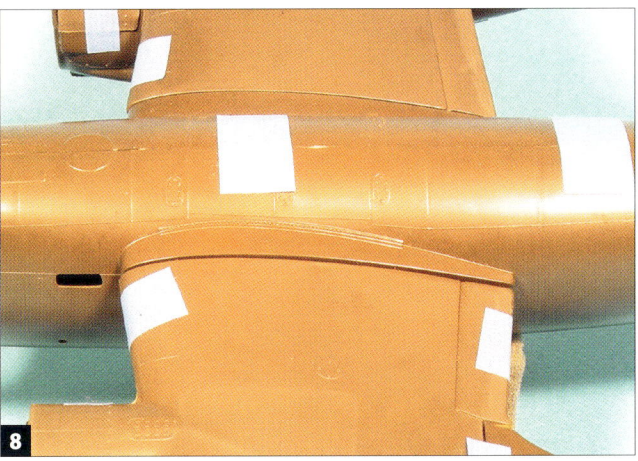

8

Photos 8-10: Filling these gaps requires a number of different filling techniques. Shims were best for the wing and tailplane seams, while filler putty worked well around the nacelles.

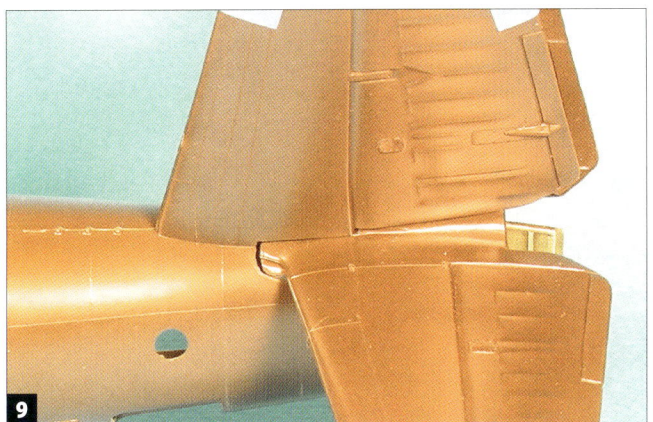

9

10

from .010" sheet styrene. I couldn't find photos of it, so I made an educated guess that this door had an accordion fold. I also added a scratchbuilt map bin from .005" and .010" sheet styrene on the back of the pilot's seat, **6**.

Another new item was a bulkhead from .010" sheet styrene to separate the bombardier and pilot compartments, **7**. I added a storage bin and ribbing from plastic scrap to it. The first-aid kit was shaped from a bit of Sculpey modeling clay and baked.

Eliminate the negative. Having dealt with the best the kit has to offer, the next step was to overcome its biggest drawback – the fit. The initial test-fitting revealed substantial gaps where the wings, **8**, and tailplanes, **9**, met the fuselage, and the fit of the engine nacelles to the wings was also poor, **10**.

After years of debating the pros and cons of fillers, I've concluded that none is perfect. So I use several, depending on

the situation. Super glue is ready to sand minutes after application, so I use it as much as possible. As soon as the accelerator evaporates, I use hobby files to quickly bring the super glue even with the surrounding surface. I follow up by wet sanding with 400- and 600-grit sandpaper. Super glue becomes too hard to sand after an hour, so I had to work quickly.

In areas with a lot of surrounding detail that needs to be preserved or areas that are hard to reach with sandpaper, I use 3M Acryl Blue automotive filler putty. It doesn't shrink much and sands easily without much pitting.

In this case, Acryl Blue filled the seams between the nacelles and the wings. Masking tape applied to either side of the seam protected the surface detail and minimized the amount of putty that needed to be removed. Once the putty dried, I sanded with 400- and 600-grit paper to smooth everything.

It's best to apply primer over putty to fill or reveal tiny imperfections. I use auto-

body touchup paint tubes for this. They come in red or gray primer with a built-in applicator brush and are available at most automotive-supply stores. I simply paint a coat of primer over the sanded putty. A wet-sanding with 600-grit sandpaper prepares the model for paint.

In some areas, shimming large openings in joints with sheet plastic is the best solution, and that's what I did with the wing/fuselage joints. Although Monogram provides spars for the wings to slip over, there is not a lot of gluing surface. I used five-minute epoxy on the spars and held the wings in the proper dihedral with tape until the epoxy set.

To determine the shape of the shims, I inserted a sheet of .005" styrene as far as it would go into the top and bottom seams and used a pencil to mark the contour of the wing. I removed the sheet and cut away the excess using the pencil marks as a guide, leaving a custom-fit shim. I glued the styrene shims in place with liquid cement, **11**. Next came a smear of Acryl

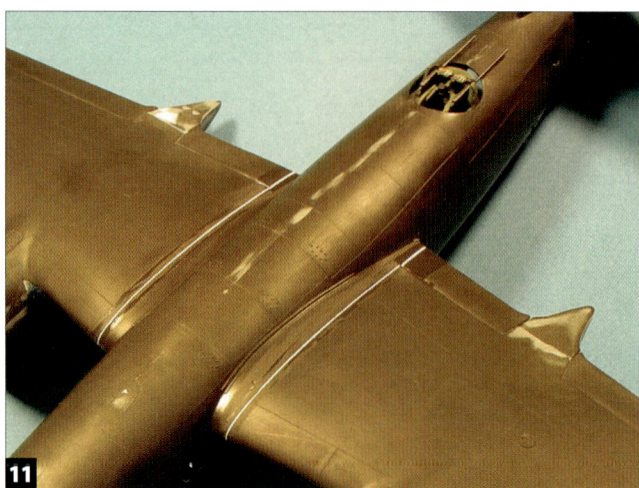

11 Rafe inserted thin sheet plastic strips into the top and bottom of the wing-to-fuselage seams.

12 A layer of putty eliminated all traces of the seam once it was sanded away.

13 All of this putty and shim work will disappear under a coat of paint. Auto primer helps fill in low spots.

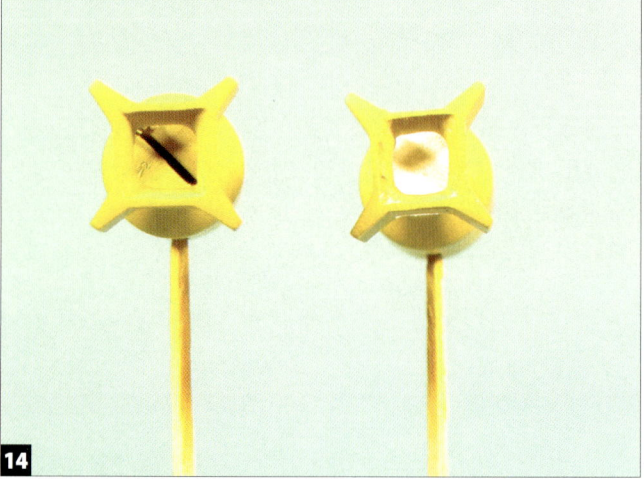

14 Rafe filled the gaps in the rear of the bombs with spackle and smoothed it with sandpaper wrapped around a toothpick.

Blue putty reduced with lacquer thinner, **12**. When that had dried overnight, I wet-sanded, brushed auto primer to fill pin-holes in the putty, and wet-sanded again with 600-grit sandpaper once the primer had dried, **13**.

Some gaps are difficult to fill and almost impossible to sand, such as the back ends of the kit bombs, **14**. My solution was to fill them with wall-board spackle. This house-hold patching material is easy to work when wet. I pushed it in and shaved off the excess with a flat toothpick, then wrapped a sliver of 600-grit sandpaper around the toothpick to do a little smoothing once the spackle had dried.

Latch on to the affirmative. The paint scheme for the early Marauders was simple – olive drab on top, neutral gray on bottom. I used Testor Model Master

Acryl for both colors, but first I applied an overall coat of light gray primer to help spot remaining blemishes. Next, I used the artist's blending stump to pre-shade all the panel lines with black pastel. Then I rubbed the pastel into the paint with a paper towel, wiping in the direction of the airflow. This step provides a little tone variation when the final colors are applied, **15**.

Once the top and bottom colors had been applied, I airbrushed olive drab lightened with a little yellow and white into the centers of a few topside panels to create visual variety. A thin coat of a light buff (made by tinting flat white with yellow and olive drab) was airbrushed over the entire top surface of the model to create a faded, weathered effect.

Finally, I applied a mixture of silver and burnt umber paint with a fine brush

in random areas along the edges of the wings and tail planes to simulate wear. I also picked out worn areas along the top of the fuselage with a Prismacolor silver pencil, **16**. Pastel chalks darken random panel lines and replicate oil streaks and exhaust stains around the nacelles, **17**. With all the weathering and paint chipping applied, the B-26 really looked as though it had seen a lot of action!

Chatter away. You might wonder about the markings – don't bother looking for references. "Chatter Box" is my own invention. I found a classic pinup, reduced it, and made a custom decal for my model.

Armed with solutions to the fit problems and intermediate modeling skills, modelers who build this kit may be just as surprised at their great results as the pilots who flew the real B-26.

15 Dark pastel dust was burnished over the panel lines to pre-shade them.

16 A light coat of buff tones down the worn-paint effect. Rafe mixed one part buff to nine parts thinner for the overspray.

MARAUDER OR MURDERER?

The Martin B-26 has to be a contender for the best-looking medium bomber of World War II. Despite its "hundred miles an hour sitting still" looks, many fledgling U.S. pilots were reluctant, if not outright afraid, to fly the B-26 due to its initial high accident rate. "One a day in Tampa Bay" became the fatalistic catch phrase for the training crews at MacDill Field in Florida, and for a while, it came close to accurately reflecting the loss rate.

Investigations revealed that the large number of trainee crashes were not the result of any intrinsic flaw in the airplane but rather that it was a hot ship that demanded the most from a talented pilot. The relatively small wing necessitated higher-than-normal landing speeds that could easily get the best of a young and inexperienced aviator. In addition, this was one of the few cases in which the U.S. Army Air Corps ordered a plane straight from the drawing board with no test prototype. As a result, a number of minor modifications were needed to realize the potential of the design.

Eventually, Lt. Col. James Doolittle, leader of the famous raid on Tokyo, was sent to MacDill to rescue the program from cancellation. Told by the student pilots that an engine failure on take-off was guaranteed to be fatal, Doolittle promptly climbed into the first B-26 on the flight line. The trainees gaped in amazement as Doolittle took off and promptly shut down and feathered the prop of one engine seconds after the wheels left the ground. Racking the plane around in a graceful turn, Doolittle then brought the bomber in for a perfect single-engine landing. From that point forward, there was no question of whether the B-26 could be mastered. It went on to earn an enviable record in the war. It also held the distinction of having the lowest loss record in action of any U.S. bomber type and earned the reputation of being a real "pilot's airplane."

– Rafe Morrisey

17 With the exhaust and oil stains simulated by black and raw umber chalk pastels, the model comes to life. Rafe sealed all the weathering with a mist of clear flat.

REFERENCES

B-26 Marauder in Action Steve Birdsall, Squadron/Signal Publications, Carrollton, Texas, 1984

U.S. Bombers of World War Two Robert Dorr, Arms & Armour Press, London, 1989

Famous Bombers of World War II William Green, Doubleday & Co., New York, 1975

Deadly Duo: The B-25 and B-26 in WWII Charles Mendenhall, Specialty Press, Osceola, Wisconsin, 1981

Detailing a EUROPEAN THEATER B-24

Aftermarket parts and careful weathering liberate the full potential of Monogram's old 1/48 kit

BY GARFIELD INGRAM

The sheer size of a B-24H in 1/48 scale encouraged Garfield to enhance his model's "curb appeal" with colorful markings, patchy panels, and extensive weathering.

Big four-engine World War II bombers are impressive in 1/48 scale, especially compared to their single-engine "little friends." After building a Lancaster and a vacuum-formed Halifax, I decided my next large-scale project had to be the good ol' B-24 Liberator. The only kit in that scale is Monogram's quarter-century-old offering – starting to show its age but basically accurate. This diamond in the rough just needed some extra detailing and correcting to become a real gem.

Major-component construction followed the instructions, so I'll focus on specific details I added for improvement. Before I cut plastic, however, I planned the process to avoid adding details that wouldn't be visible.

Interior design. The waist-gun compartment is clearly visible through the open windows, so I needed to add a lot of detail there, **1**. The walls, bulkheads, and floors were painted interior green, then lightly weathered in traffic areas with dry-brushed aluminum. I made the plywood panels lining the fuselage walls at the waist-gunner stations (probably meant to protect the skin of the aircraft from the jostling of the gunners and their equipment) from sheet styrene. I cut the pieces to size, then sprayed them with a wood-tone base color and hand-painted streaks to represent the wood grain. I added oxygen bottles, wiring, ammo chutes, and Paragon .50-caliber machine guns.

The molded-in detail of the main-gear wells is very basic, so I removed the wells from the lower wing halves, **2**, and built more accurate interiors from styrene strip, **3**. All those museum tours actually paid off! To prevent the finished model from tipping back on its tail, I filled the compartment above the new well ceilings with a mixture of lead shot and white glue, taping the wing halves together while the glue hardened. The landing gear is hefty enough to take the weight, but I made sure the parts were glued firmly together.

I painted the wells interior green before permanently sealing the wing halves, then added scratchbuilt detail and True Details wheels to the landing gear struts, **4**.

1 Clearly visible through the open windows, the waist-gun compartment needed extensive detailing, including oxygen bottles, wiring, machine guns, and simulated plywood wall guards.

2 Out goes the poorly detailed kit wheel well and …

3 … in comes a more-accurate interior made from styrene strip.

4 The wheel wells were painted interior green before the wing halves were glued. Garfield then detailed the landing-gear struts and added True Details wheels. Note the nicely weathered exhaust pipes and superchargers.

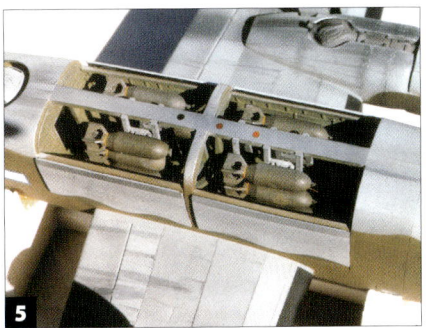

5 The most prominent detail in the full bomb bay is the cargo, so Garfield carefully weathered the bombs to look like they'd been nicked and scuffed in handling.

6 The kit's clear parts were replaced with aftermarket items from Squadron and Koster.

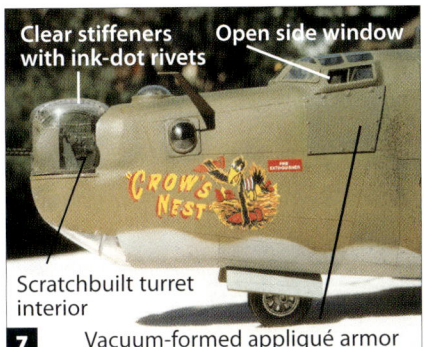

Clear stiffeners with ink-dot rivets

Open side window

Scratchbuilt turret interior

Vacuum-formed appliqué armor

7 With its clear parts meeting at exposed seams, the large nose turret was tricky to assemble.

Trailing edge of wing thinned before assembly

Opened vents

Corrugations made from angle stock

8 Garfield thinned the trailing edges of the wings to scale thickness and opened up the vents at the rear of the engine nacelles.

9 Rather than remove the control surfaces to reposition them, Garfield chose to accentuate the lines with a razor saw and scribing tool.

The upper reaches of the bomb bay are hard to see with a full bomb load in place, so the interior received minimal detailing beyond the bomb racks. I painted and weathered the bombs to give them a nicked-and-scuffed "handled" look, **5**.

Wings, windows, and turrets. With prominent clear acrylic turrets and the substantial "greenhouse" over the cockpit, the quality of the clear parts makes or breaks a B-24 model. I opted for thin, distortion-free aftermarket replacements from Koster and Squadron, **6**. For maximum clarity, I dipped each part in Future floor polish and placed it on a tissue to wick away excess liquid before drying overnight.

After making sure all the frame lines were straight, I masked Squadron's cockpit greenhouse with tape. Some modelers use liquid masking solution in this type of situation, but I find that peeling off the dried solution from the part after painting also removes the underlying coat of Future.

The turrets were tricky because there are no solid frames to hide the glue joints.

The complex clear parts had to be extensively test-fitted and trimmed before gluing with Micro Kristal Klear.

For the large nose turret, **7**, I scratch-built an interior, then assembled the clear three-piece main housing with thin strips of tape. After checking to assure fit, I glued the parts with thinned Kristal Klear. Dots of ink represent rivets in the clear stiffeners.

I attached the greenhouse with small amounts of super glue, then lightly filled and sanded the seams. Next came the side windows. I made window frames by vacuum-forming thin styrene sheet over carved forms. The frames on the aircraft I was modeling were much heavier than on a standard B-24H, perhaps to accommodate thick armor-glass side windows. I positioned one pane of the pilot's window open to show the cockpit detail. The field-installed appliqué armor under the side windows was made from sheet plastic with rivets simulated by small drops of white glue.

The wings were too thick, so I thinned the trailing edges by sanding the inner surfaces on a piece of sandpaper glued to a board, **8**. I added oil coolers to the engine compartments and opened the vents at the rear of the nacelles.

To avoid the hassle of cutting away the control surfaces to reposition them, I deepened the lines around them with a razor saw and a scribing tool, **9**. Short lengths of styrene angle stock were applied to simulate the corrugation on the vertical edges of the sliding bomb-bay doors.

Painting. The standard European theater olive-drab bomber scheme is flat as a flounder, so I selected the markings of Crow's Nest, an 8th Air Force B-24H. The colorful nose art and tail bands give the model more curb appeal, **10**. All of the painting was done with Testor Model Master paints.

Extensive decaling tends to pile up layers of clear coat, so painting began by spraying on the major markings like the distinctive divisional marking on top of the right wing (see photo **9**). I used masks cut from Frisket film with an Olfa Circle Cutter, a straightedge, and a hobby knife.

When the markings were dry, I masked them and created the correct

10 Painting involved applying several shades of the basic olive drab and gray colors. The bright, G-rated nose art and red-and-white tail markings add visual interest.

11 Four 1,200-horsepower Pratt & Whitneys kick out a lot of exhaust and oil, so the underside had to be heavily weathered.

WORKHORSE OF THE AIR WAR OVER EUROPE

Although less famous than the B-17s, the more numerous B-24s served with distinction in all World War II theaters. With more than 18,000 produced, more B-24s were built than any other U.S. warplane.

Designed by Consolidated Aircraft to meet a U.S. Army Air Corps specification, the B-24 prototype first flew in December 1939. Orders immediately poured in, first from France and Britain, then from the USAAC as the United States entered the war. Consolidated didn't have the capacity to keep up with the demand, so B-24s were also license-built by Ford and Douglas.

With a larger bomb capacity than the B-17 (8,000 lbs. vs. 6,000 lbs.) and an ability to absorb tremendous damage and still get its crew home, the B-24 became the workhorse of the 8th Air Force in Europe. Most were G, H, or J models, but there was very little difference among them (Gs and Hs could be upgraded to J standards in the field).

B-24H-20-FO (No. 42-96010) was assigned to the 466th Bomb Group and nicknamed Crow's Nest by its crew. On Oct. 25, 1944, commanded by Lt. C. Maxton, it departed an airfield near Norwich, England, on a mission to attack synthetic fuel plants in Germany near Gelsenkirchen. Over the target, it was heavily damaged by anti-aircraft fire and crash-landed near Brussels.

– Lawrence Hansen

patchy look on the top surfaces by applying six different shades of olive drab. The underside received varying shades of neutral gray to replicate a worn, mud-spattered, oil-stained appearance, **11**. I accentuated the patchy effect by highlighting areas like the cockpit side armor plates with full-strength paint to simulate field touch-ups.

Finishing and decaling. I enhanced the weathering with several washes of thinned acrylics and added highlights with artist's pastel chalk and Prismacolor pencils. It sounds like a lot of work, but when I was done, I wondered if I hadn't taken it far enough!

I had selected a prototype with colorful nose art suitable for model-show

viewers of all ages. I made the decal by enlarging a photograph to 1/48 scale on a photocopier, painting it with opaque watercolors, and finally color-copying it onto decal paper.

At least three B-24s in the 8th Air Force bore the Crow's Nest name and similar nose art. However, available photos of "my" ship don't show the rear of the aircraft, so I had to guess at the ID letter on the vertical stabilizers.

A project of this size is a long haul, but researching and detailing this unique aircraft were a lot of fun. Until some high-tech manufacturer comes out with a new, state-of-the-art B-24 in this scale, Monogram's old chestnut can still serve as the basis of an impressive model.

Painting a
NORTH AFRICAN "EMIL"

Finishing Tamiya's 1/48 scale Bf 109E-4 as "Black 8" STORY AND PHOTOS BY KOSTAS DIMITROPOULOS

One of the more spectacular Luftwaffe camouflage schemes was applied to the first Messerschmitts in North Africa. Kostas Dimitropoulos shows us how to paint the Tamiya Bf 109 kit.

I still remember the multi-colored Matchbox kits I built when I started in the hobby back in the 1970s. Recent kits are exquisitely detailed, fit well, and usually need nothing more than careful assembly and a nice paint job to look like the real thing.

Tamiya's 1/48 scale Messerschmitt Bf-109E-4/7 falls firmly into this category: a kit with almost perfect fit and fine detail inside and out! It features recessed panel lines and separately molded slats, flaps, and rudder that allow the model to be built in more realistic poses. A complete set of decals provides markings for three machines, including the one I chose, a beautifully camouflaged E-4 of 2./JG 27 that flew in North Africa.

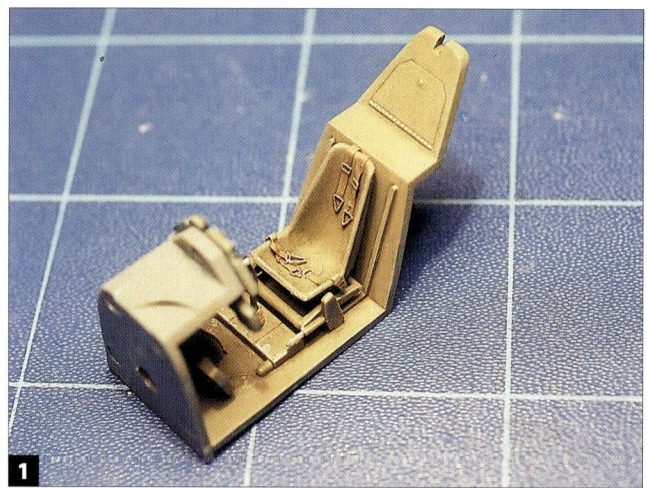

1 Instead of using the kit decal harness, Kostas cut lead foil for the straps and attached photoetched metal parts for the buckles.

2 A thin dark wash accents recesses around cockpit details.

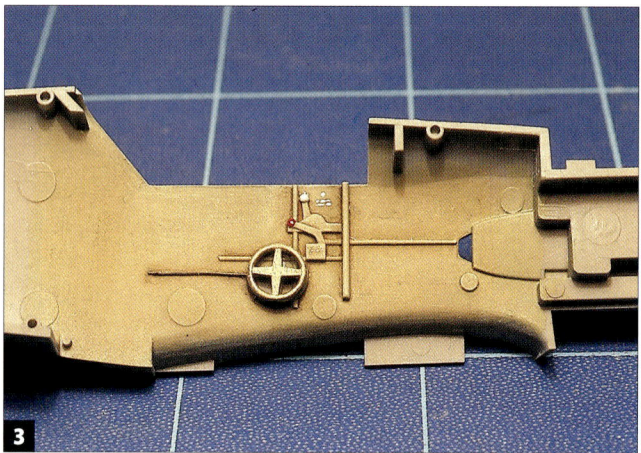

3 Dry-brushing adds visual depth to the interior by highlighting raised details.

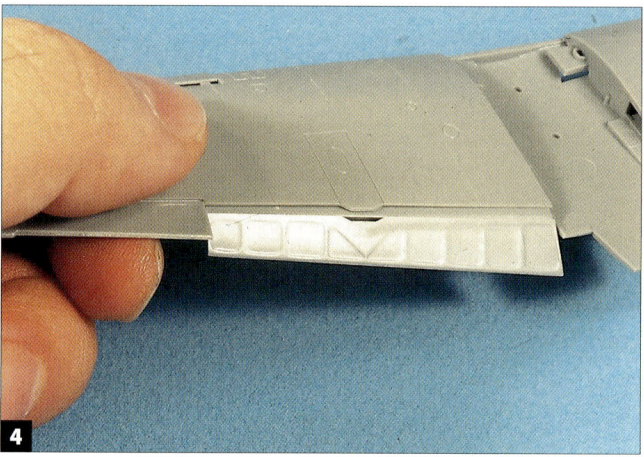

4 Tamiya's kit allowed Kostas to drop the flaps and slats for a realistic appearance.

Getting started. The cockpit comes with nice raised details on the instrument panel and the side consoles. There is also a decal for the instrument panel that will prove useful for those who don't trust their painting skills, but I opted for painting the knobs and various gauges with a pointed 5-0 red sable brush. Of course, that was done after adding some minor details on the side walls, such as the oxygen hose and wiring from thin copper wire.

I improved the seat with lead foil harnesses and Reheat Models' photoetched brass buckles, **1**. I painted the cockpit RLM 02 gray (Humbrol No. 31) and the oxygen bottle with Testor steel. The harnesses and the headrest were painted leather (Humbrol No. 62) and the instrument panel black.

I used RLM 66 dark gray for the interior canopy framing and replaced the gunsight reflector with thin clear plastic. The whole cockpit assembly, including the side walls, received an overspray of Winsor & Newton's matte acrylic clear varnish (available in art-supply stores). I helped details stand out with a wash of Talens' raw umber oil paint, **2**.

When the wash was set, I dry-brushed a mixture of RLM 02 and white, **3**. I picked out knobs and switches with enamels, then represented the instrument glass covers with drops of 5-minute epoxy. I decided to install the cockpit tub up through the opening in the assembled fuselage, rather than glue it to one fuselage side. This ensured perfect alignment.

The single bottom and two top portions of the wing help establish the proper dihedral. I glued the separate slats and flaps in the dropped position, typical of Messerschmitts on the ground, **4**. A piece of sheet styrene inside prevents seeing

through the radiator, **5**. Next, I brushed liquid cement into the wing/fuselage joint. I needed only a little filler on the lower rear joint, **6**.

The undercarriage legs were improved with copper wire brake lines and were painted RLM 02 along with the wheels, doors, and wheel wells, **7**. They were weathered with a thin wash of black oil.

That deserted look. The Luftwaffe's JG 27 was the first fighter unit to arrive in North Africa, preparing for operations in Libya in April 1941. Being big advocates of camouflage, the Germans wasted no time in painting their Bf 109s in an effective scheme of RLM 78 Hellblau (light blue) for the bottom and sides and RLM 79 Sandgelb (sand yellow) mottled with RLM 80 Olivgrün (olive green) topsides.

Before applying the camouflage, I carefully attached the windscreen and the aft

5 A piece of sheet styrene prevents that "see-through" look in the nose radiator.

6 A little filler putty was needed at the back of the ventral wing/fuselage joint.

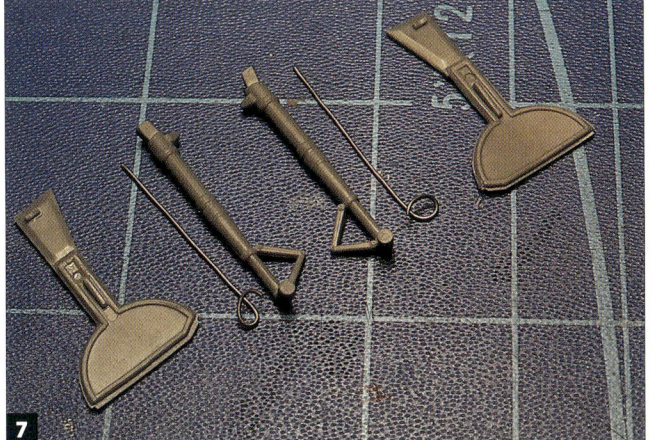

7 Copper wire added to the landing gear struts simulates hydraulic brake lines.

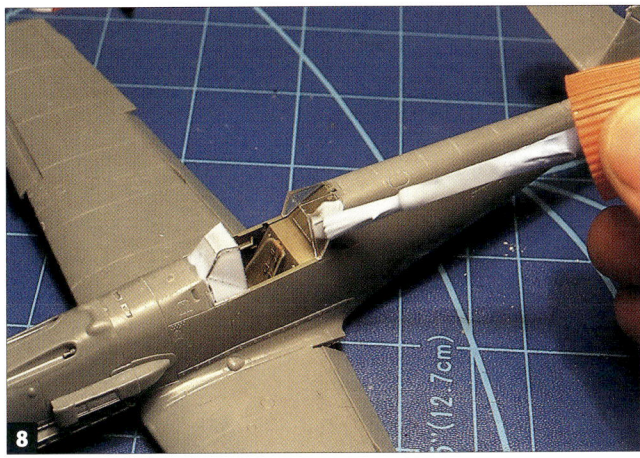

8 Liquid masking agent covers the canopy.

canopy glazing with liquid cement and filled gaps around the clear parts with white glue. Even when dry, excess white glue could be removed with a damp cloth. I masked the cockpit with Tamiya tape, then used Gunze Sangyo's Mr. Masking Sol, an excellent liquid masker, on the canopy, **8**. It's applied with the provided brush, allowed to dry, and cut away from the frames with a sharp blade.

To handle the model easily during painting, I inserted old tweezers in the propeller opening, **9**. I painted most of the model first with the lightest color, RLM 78 (Gunze Sangyo Aqueous H418) thinned with Tamiya's acrylic thinner and applied with my trusty Badger 175 Crescendo double-action airbrush. This color served as a primer to make it easier to spot flaws. I fixed a few with Gunze Sangyo's Mr. Surfacer 1000 (liquid putty) and sanded them smooth.

I resumed painting with RLM 79 (Gunze H66) lightened with 10 percent white, **10**. I faded it a little by spraying tan along panel lines on the top surfaces where the hot desert sun bleached colors considerably, **11**. Subtlety is key here.

To paint the RLM 80 (Gunze H420) mottle, I switched to my fine-tipped Badger 150 and regulated the air pressure to attain sharp edges around each patch. I carefully sprayed random faded points inside each patch with a well-thinned mixture of two parts RLM 80 and one part white.

The white ID band on the aft fuselage is included in the decals, but I chose to paint it for better results. I masked it with Tamiya tape, **12**. I also masked and airbrushed the red-and-white spinner, **13**, then used Gunze H65 black green on the prop blades.

When all colors had completely dried, I sprayed the model with Johnson's Klear

(the European equivalent of Future floor polish), to prepare the surfaces for decaling. I used the kit-provided decals for "Black 8," a 2./JG 27 machine flown by Lt. Werner Schroer during 1941. Although a little on the thick side, they conformed perfectly with the help of Gunze's Mr. Mark Softer, **14**.

Almost there. When the model was dry, I overcoated it with Winsor & Newton matte acrylic clear varnish. This protected the camouflage from the weathering abuse that followed.

Next up was a wash of raw-umber oil paint (diluted 1:8 with mineral spirits), applied to all recessed lines and corners. After it dried for about 10 minutes, I wiped away the excess with a cotton swab moistened with mineral spirits, **15**.

I airbrushed the exhaust stains with a mixture of flat black and flat red paint. Pas-

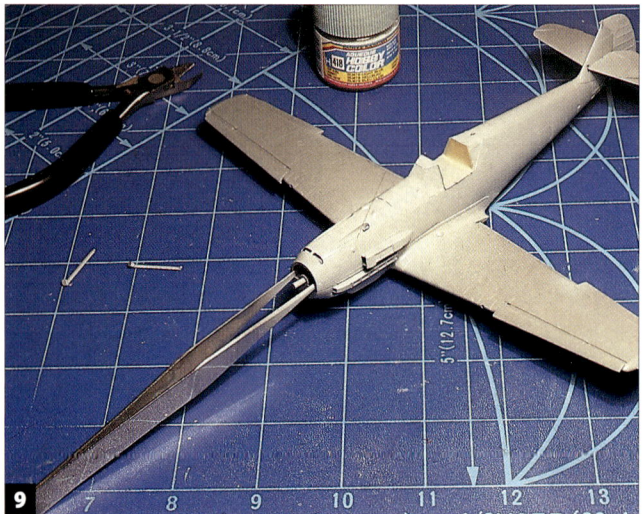

9 To make it easier to paint the model, Kostas used tweezers as a makeshift handle.

10 Kostas applies the topside desert yellow with an airbrush.

11 Tan applied to panel lines gives a faded appearance to the upper surfaces.

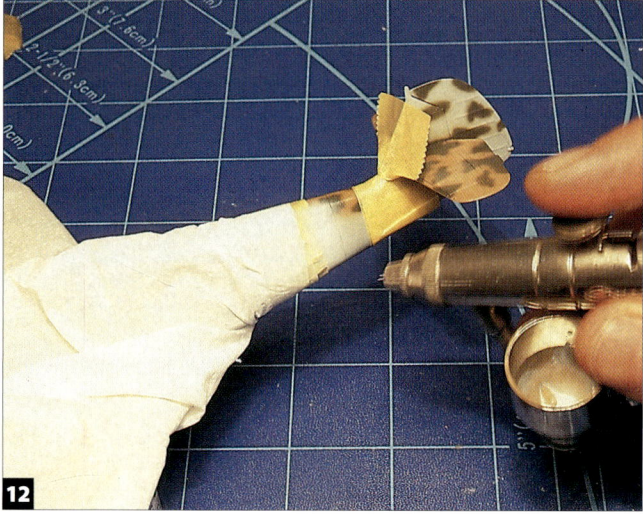

12 Kostas airbrushed the ID band on the rear fuselage.

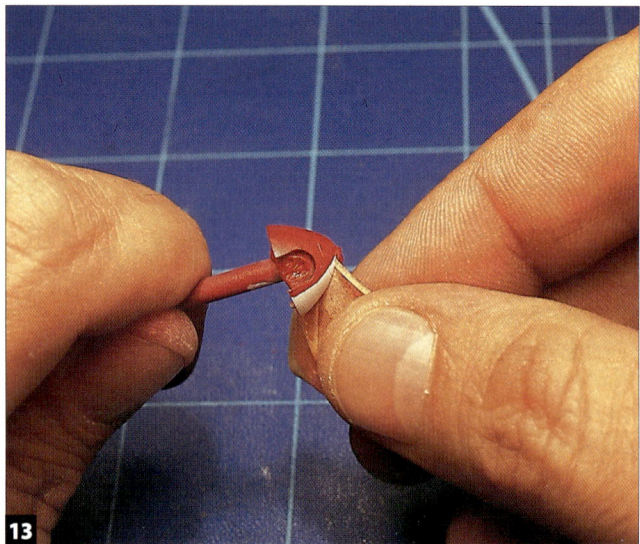

13 The spinner was painted white first, then half of it was masked and painted red.

14 A setting solution helps the decals snuggle down into the recessed detail.

15 After sealing the model with a varnish, Kostas accented panel lines with a wash of thinned oil paint.

REFERENCES

Building the Messerschmitt Bf 109 Kevin Hjermstad and Glen Phillips, Kalmbach Books, Waukesha, Wisconsin, 1998

Wings of Fame Vol. 4 Aerospace Publishing, London, England, 1996

Bf 109 in Action (Part I) Squadron/Signal Publications, Carrollton, Texas

Bf 109 Aces of North Africa and the Mediterranean Jerry Scutts, Osprey Publishing, London, England, 1997

Messerschmitt Bf 109A-E Willy Radinger and Walter Schick, Schiffer Publications, Atglen, Pennsylvania, 1999

Kostas' finished model is a handsome addition to his WWII fighter collection.

tels took care of the fading of insignia and the soot around the guns. I painted the gun barrels with a mixture of Humbrol Metalcote polished steel and polished aluminum.

A few smears of thinned raw umber and black oil paints around the wheel wells and spinner, and a little scuffing with aluminum in selected panel lines, walkways, and filler caps added the final touches.

The last steps included gluing the canopy in the open position and adding a stretched-sprue radio antenna. Aerial insulators were made from dabs of 5-minute epoxy and painted with polished steel. I painted the navigation lights silver, then overcoated with Tamiya acrylic clear green on the right and clear red on the left.

Well, that's it – an easy-to-build Messerschmitt Bf 109 "Emil," with its desert camouflage – a must-have piece in any WWII collection.

Modeling a late-war
MESSERSCHMITT

Fine-tuning Revell-Monogram's 1/48 scale Bf 109G BY JIM GREEN

Messerschmitt Bf 109Gs soldiered on until the end of World War II, sometimes in unusual camouflage schemes like this one.

F ew would dispute that Willy Messerschmitt's Bf 109 was synonymous with Germany during World War II. Its development mirrored the Reich's changing fortunes as the war progressed: from record-breaking thoroughbred in the late 1930s to overburdened war horse in 1945.

In its later variants it was required to keep up with more advanced Allied designs, and because suitable replacements weren't available in substantial numbers, it was kept in front-line service

even as defeat embraced the Third Reich.

Most Bf 109 models that I've seen have represented aircraft flown by well-known aces (such as Erich Hartmann) or planes from famous fighter units. I decided to build one that would be different – typical of the late-war period, and characterizing the weariness of the Luftwaffe's struggle during the war's last days.

Revell-Monogram's kit was tooled in the late 1970s. Its detail may not rival the best of today's kits, but it's a good late Bf-109G-10. The kit's only shortcomings

are a cockpit and landing gear that lack detail. Both areas can be remedied easily with aftermarket items, a few spare parts, and a little creativity.

Cockpit improvements. First, I corrected and detailed the kit's instrument panel and the cockpit floor as outlined in drawing **1**. I used True Details' photoetched set No. 48812 along with instrument decals from ProModeler's German fighter instrument sheet. The end result is a vast improvement, **2**. I covered the instru-

INTERIOR IMPROVEMENTS

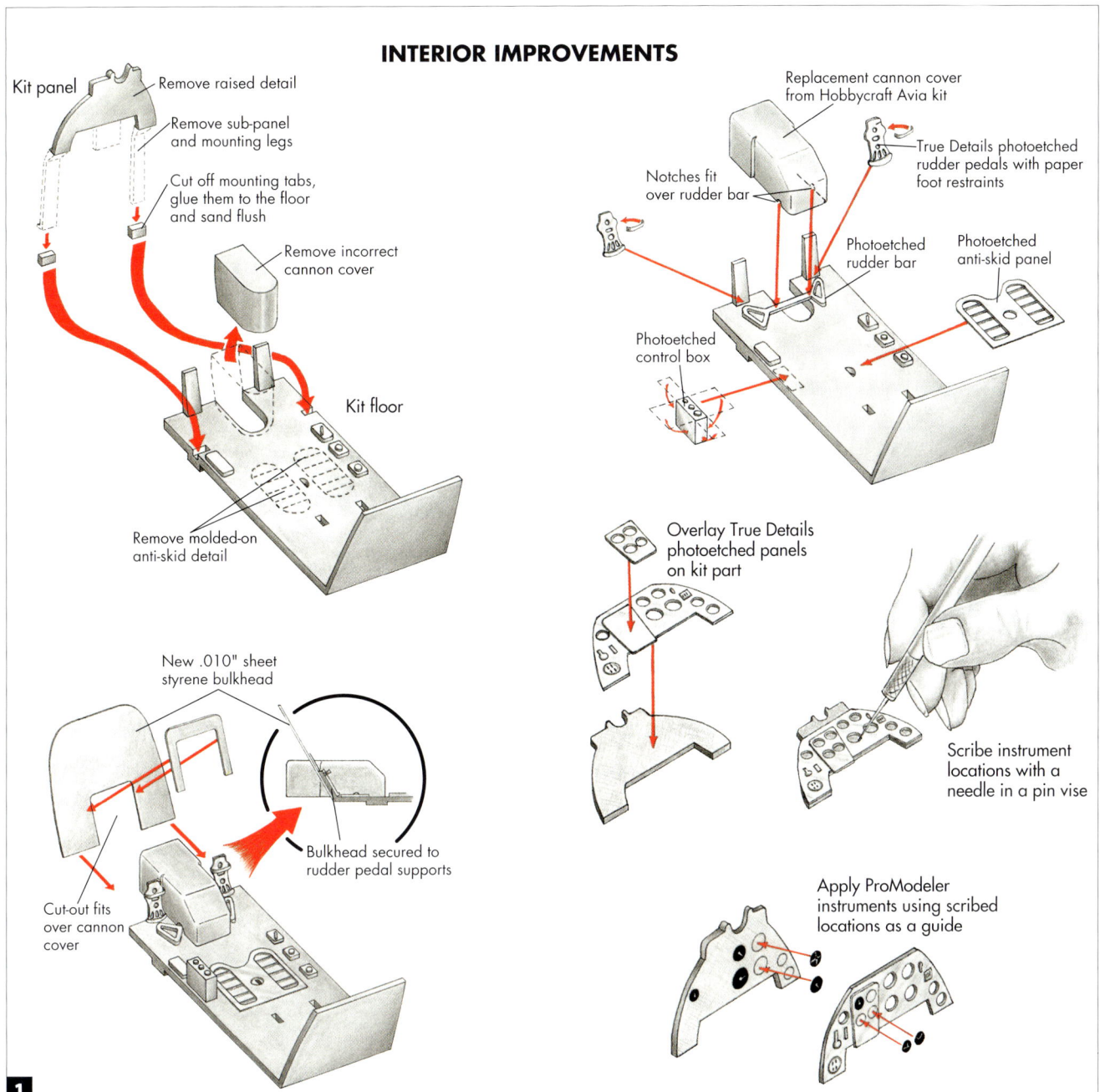

Kit panel — Remove raised detail

Remove sub-panel and mounting legs

Cut off mounting tabs, glue them to the floor and sand flush

Remove incorrect cannon cover

Kit floor

Remove molded-on anti-skid detail

New .010" sheet styrene bulkhead

Bulkhead secured to rudder pedal supports

Cut-out fits over cannon cover

Replacement cannon cover from Hobbycraft Avia kit

True Details photoetched rudder pedals with paper foot restraints

Notches fit over rudder bar

Photoetched rudder bar

Photoetched anti-skid panel

Photoetched control box

Overlay True Details photoetched panels on kit part

Scribe instrument locations with a needle in a pin vise

Apply ProModeler instruments using scribed locations as a guide

1

---

ments with drops of white glue, which dries clear and gives the impression of a glass cover.

Next, I detailed the cockpit sidewalls, **3** and **4**. Removing the raised details made way for the new parts. Grooves scored into both fuselage halves accept the rebuilt instrument panel. I have found pilot's accessory catalogs such as Sporty's to be a great source of photos of instruments and placards. I can usually find some that will fit any project.

At this time, I mounted the kit's exhaust stacks and the oil cooler with the photoetched intake screen. I improved the

exhaust pipes by making several passes with a knife-edged file to separate the stacks, then drilled them out with a small bit mounted in a pin vise.

Next, I applied the photoetched lap belts to the pilot's seat and set that assembly aside for later. The kit's gunsight was replaced with a pair of reflectors made from thin acetate, also set aside for later.

I painted the interior with Testor Model Master enamels. I airbrushed flat black on the floorboard and side walls, then dry-brushed with Euro I dark gray. Individual items were then picked out in appropriate colors: yellow for the fuel line, insignia

blue for the oxygen regulator, dry-brushed silver for the rudder pedals and bar, and brown for the stick handle and pedal foot restraints. I weathered the finish using brown watercolors and pastels, then picked out worn and scuffed areas with a silver pencil (found in art supply stores). An airbrushed application of Dullcote was the final touch.

Canopy improvements. I had planned from the start to add a Squadron vacuum-formed "Galland hood" canopy to model a late G-10 variant. I also wanted to correct the shape of the cockpit opening.

2

The new interior tub looks great when painted.

Labels: Photoetched details in place; New bulkhead blanks out forward fuselage

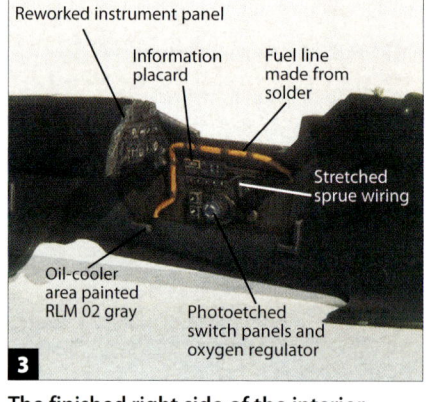

3

The finished right side of the interior shows the characteristic yellow fuel line.

Labels: Reworked instrument panel; Information placard; Fuel line made from solder; Stretched sprue wiring; Oil-cooler area painted RLM 02 gray; Photoetched switch panels and oxygen regulator

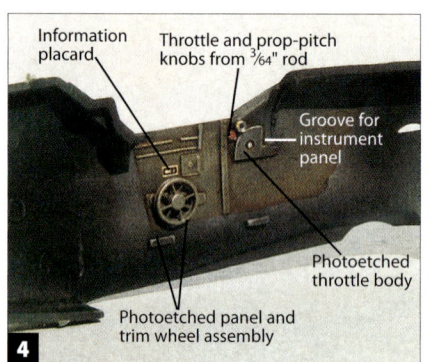

4

The left side of the interior shows throttle and controls.

Labels: Information placard; Throttle and prop-pitch knobs from ³⁄₆₄" rod; Groove for instrument panel; Photoetched throttle body; Photoetched panel and trim wheel assembly

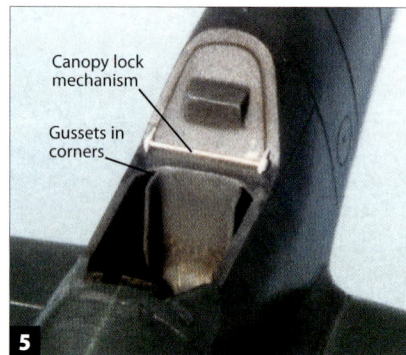

5

Here the seat has been blended to the deck and the canopy lock mechanism has been added from styrene sheet and wire.

Labels: Canopy lock mechanism; Gussets in corners

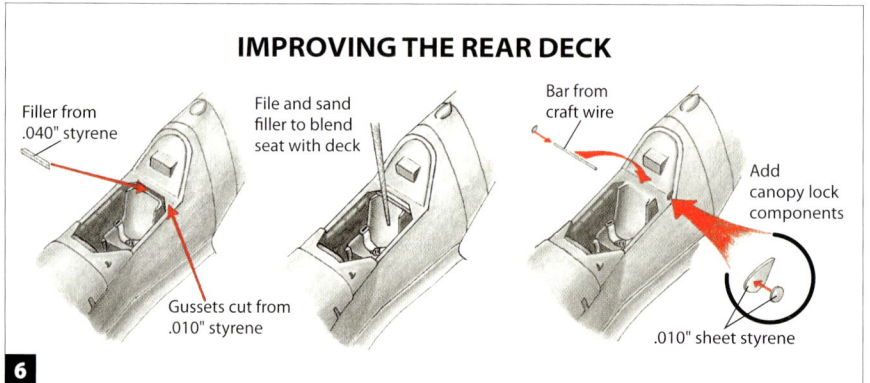

IMPROVING THE REAR DECK

6

Labels: Filler from .040" styrene; File and sand filler to blend seat with deck; Bar from craft wire; Gussets cut from .010" styrene; Add canopy lock components; .010" sheet styrene

7

The kit windscreen received improvements. Note the dangling harness strap.

Labels: Photoetched rear frame; Improved gunsight in place

8

The photoetched seat harness was bent to shape.

Labels: Margarine foil bracket; Left-side harness added after the model was painted

After assembling the fuselage and mating the wings and tailplane, I filled the gap between the back of the kit's floorboard and the underside of the cockpit decking with a shim of plastic and super glue. I smoothed the joint with a homemade tool of sandpaper strips wrapped around a section of Plastruct I-beam stock, held by angled cross-locking tweezers.

To fashion the proper shape of the cockpit opening, I added .010" sheet triangular gussets to the rear corners and blended them in, **5**. In preparation for painting the back deck, I used an airbrush (with no paint, obviously) to blast sanding dust from the cockpit, then cut drafting tape to size and masked the rear bulkhead below the repaired area. I protected the rest of the cockpit by stuffing in wet tissue and painted the rear deck with flat black, followed with dry-brushed Euro I dark gray.

After unmasking the cockpit, I installed the seat, only to discover the inaccurate step between the top of the seat and the back deck. Since this configuration wouldn't allow the photoetched shoulder harnesses to "hang" properly, I corrected it with a sheet-styrene shim and filed it smooth. I also added canopy-lock detail from bits of styrene and wire, **6**. The retaining bracket was made from a piece of foil margarine wrapper attached with white glue.

Next, I thinned the inside edge of the kit's windscreen with a knife to accept the photoetched windshield frame, **7**. I wet-sanded and polished the area smooth, cleaned up the residue with a cotton swab and denatured alcohol, then attached the windshield frame with super glue. I mounted the gunsight to the instrument panel, then glued the windscreen to the model. A light sanding blended the rear frame into the windscreen.

After final painting to the cockpit deck, I added the right shoulder harness and pre-bent (but did not install) the left one, **8**. I wanted to drape the left harness over the cockpit sill, so I added it after painting the airframe.

I modified the kit's canopy head armor so it would fit inside Squadron's vacuum-formed canopy, **9**. The kit part is designed to fit the thicker injection-molded canopy, so I stole new mounting brackets from a 1/72 scale German assault gun that were wide enough to fit inside the thin vacuum-formed hood, **10**. The armor frame was painted RLM black-gray and attached to the inside of the hood with white glue.

Finished with the cockpit, I masked it and the glass panels of the windscreen and canopy and airbrushed the interior frame color of black-gray.

Exterior enhancements. The kit's gun barrels lacked detail, so I cut them out and deepened the troughs with a needle file, **11**. I found a pair of gun barrels in my spares box and carefully bored out the muzzles with a fine drill bit.

The next item up for the detailer's treatment was the supercharger intake, **12**. After gluing the intake halves together, I filed off the locating tab and bored a hole in the flat side that faces the fuselage. I also bored out the intake with a ball-shaped cutting bit turned between my thumb and forefinger. I used a Waldron punch set to make a screen disc and inserted it into the hole in the flat side. I painted the inside of the intake with RLM 02 gray.

Using a round file, I notched out the wing-tip navigation lights, then cut new ones from the corners of a .040" clear sheet. These were filed and sanded to fit in the notches. I used a fine drill bit to bore a depression in the back side of each lens and applied a tiny drop of red (left) and green paint to simulate the bulb. I attached the new lenses to the notch with gel-type super glue, then sanded and polished them to clarity, **13**.

Painting. I painted my model as a Reich Defense machine of JG 27 using Testor Model Master enamels. The scheme was RLM 75 gray-violet and RLM 81 brown-violet topsides over RLM 76 light blue. Testor RLM 81 looked more like Olive Drab, so I used French Chestnut Brown instead.

Before applying the camouflage, though, I painted the wheel wells RLM 02 gray, then painted the nose stripe yellow and the Reich Defense tail band green. Next I painted the area of the upper wing crosses white. Masks cut from sign-maker's "Gerber mask" were laid on the white to create the white cross outlines. These stayed on during the application of the camouflage. When all the painting was done, the masks were removed to reveal the white-outlined cross.

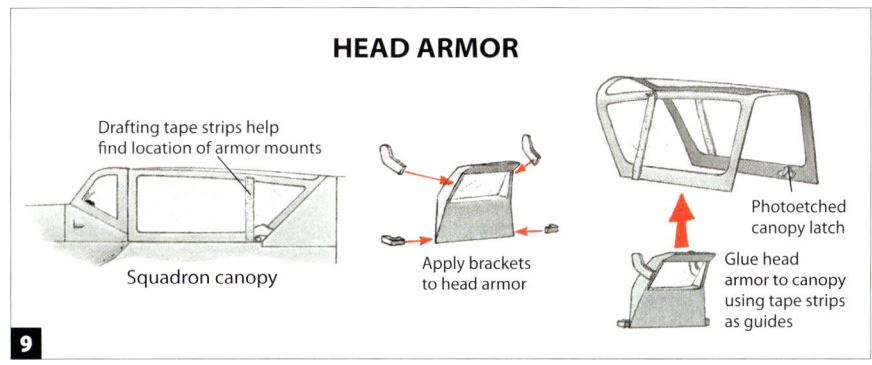

HEAD ARMOR

Drafting tape strips help find location of armor mounts

Squadron canopy

Apply brackets to head armor

Photoetched canopy latch

Glue head armor to canopy using tape strips as guides

9

10

The finished Galland hood includes the modified head armor.

Molded-on gun barrels removed and troughs deepened with a needle file

Replacement barrels from spares box

11

The nose guns were poorly defined in the kit moldings, so improvements helped.

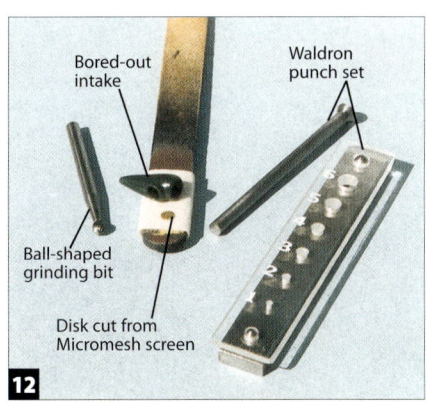

Bored-out intake

Waldron punch set

Ball-shaped grinding bit

Disk cut from Micromesh screen

12

The small supercharger inlet was drilled out and refurbished.

13

New wing-tip navigation lights were made from clear plastic.

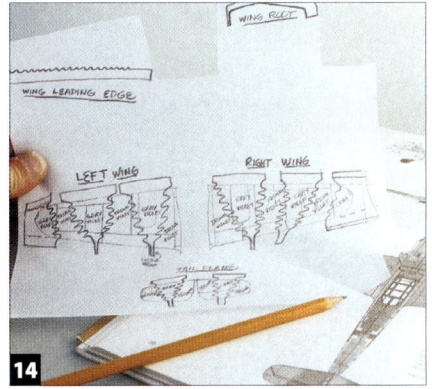

14

Drawings of the camouflage scheme were traced onto card stock and used as soft masks.

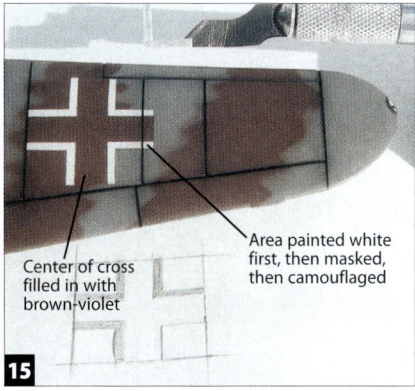

Center of cross filled in with brown-violet

Area painted white first, then masked, then camouflaged

15

On this model, the upper-surface crosses are not decals. Jim painted the white outlines first, then filled in the centers.

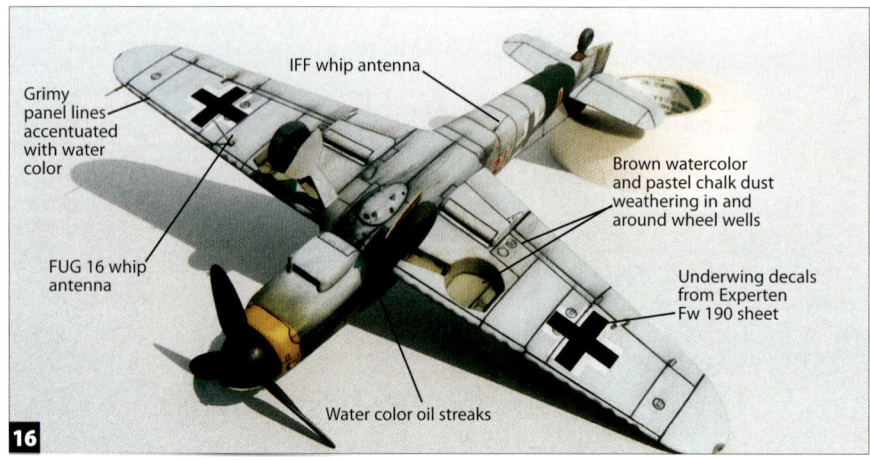

Grimy panel lines accentuated with water color

IFF whip antenna

Brown watercolor and pastel chalk dust weathering in and around wheel wells

FUG 16 whip antenna

Underwing decals from Experten Fw 190 sheet

Water color oil streaks

16

The weathering on the bottom of the model helps make it look war-weary.

Three Guys swastika

Experten serial number

400240

Clear light from drop of white glue

17

The tail end shows the mottled camouflage of the late-war scheme.

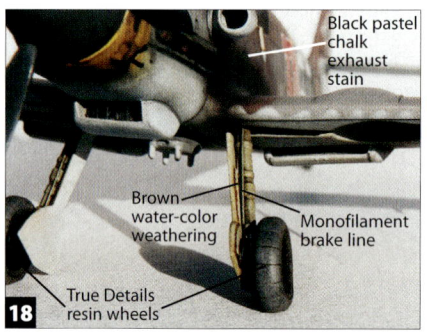

Black pastel chalk exhaust stain

Brown water-color weathering

Monofilament brake line

True Details resin wheels

18

Landing gear improvements include new wheels.

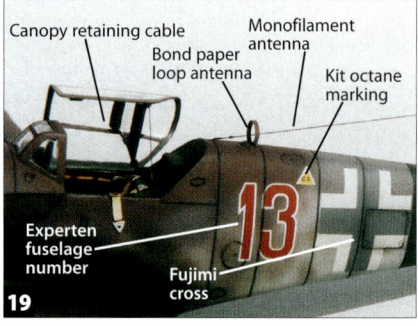

Canopy retaining cable

Monofilament antenna

Bond paper loop antenna

Kit octane marking

Experten fuselage number

Fujimi cross

19

Topside details include radio antennas.

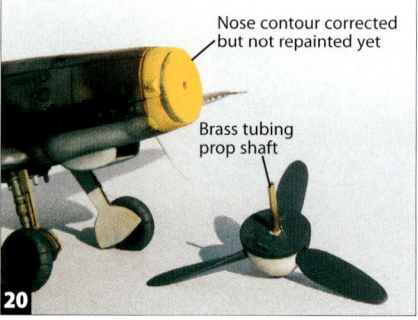

Nose contour corrected but not repainted yet

Brass tubing prop shaft

20

During final assembly, Jim found the nose was malformed.

REFERENCES

German Fighters of World War II
Franco Ragni, Squadron/Signal Publications, Carrollton, Texas, 1979

Messerschmitt Bf 109 in Action Part 2
John Beaman, Squadron/Signal Publications, 1983

Wings of Fame Vol. 11, Messerschmitt Bf 109 Part 2: The Later Variants
David Donald, Airtime Publishing, Westport, Connecticut, 1998

Warplanes of the Third Reich William Green, Doubleday and Co., New York, New York, 1970

Messerschmitt Bf 109 Robert Grinsell, Wing and Anchor Press, New York, New York, 1980

The first camouflage color to go on was RLM 76 light blue. This served as a primer and made it easier to find surface flaws. The next color was the upper surface RLM 75 gray-violet. The camouflage on the fuselage was not applied to any official pattern and seemed random. I airbrushed RLM 81 brown-violet freehand, extending the dark colors well down the fuselage sides. Each swatch of color ended in a loose, random mottle.

The wing pattern on this machine was more organized, though, with a soft, wavy sawtooth pattern. I drew scale paper patterns, **14**, cut them out, and traced them onto manila folder card stock. This stiffer material was perfect for making soft masks for the scheme.

The segments of the soft masks were applied to the model with small loops of masking tape to hold them in position and to hold them just off the surface. With the edges of the masks held away from the surface, the airbrush created a definite, yet slightly soft, edge to the pattern, **15**.

I masked around the already-masked white outlines of the upper-surface crosses and filled in with RLM 81.

Decaling and weathering. Color photos of late-war Messerschmitts reveal grime and wear from hard use, inadequate maintenance, and primitive field conditions.

First, I airbrushed Testor Dullcote over the entire model. I enhanced the panel lines with a No. 2 pencil, then dragged a bristle brush over each panel line, smudging the graphite to blend and give a grimy appearance. More Dullcote sealed the graphite to the model. I added another layer of heavy grime to the panel lines using watercolors applied with a fine-point brush. Brown was used in the wheel wells and landing gear; a mixture of brown with a little black for the undersurfaces, **16**; and black on the upper surface of the wing and for the panel lines around the engine and in the recesses around the exhaust stacks. The application of black is heaviest at the front of the aircraft.

Once the watercolors were dry, I moistened a cotton swab and wicked away excess watercolor from the panel lines, softening and blending the weathering. Dragging the cotton swab in the direction of the aircraft's slipstream prevents unnatural looking smudges of color.

Another application of Dullcote sealed the weathering. Oil streaks behind the oil cooler were made with undiluted black watercolor and again sealed with Dullcote.

Since decals adhere best to glossy surfaces, I airbrushed Testor Glosscote in the areas that would receive decals. All the decals came from my spares box. The underwing insignia, red fuselage number, and serials came from Experten's Fw 190D sheet; the fuselage crosses with gray centers and the gruppe bar both came from an old Fujimi Bf 109 kit; the swastikas are from the out-of-print Three Guys Replicas sheet, **17**; and the fuel octane symbol is from the Monogram sheet. Once the decals were dry, yet another application of Dullcote restored the flat finish.

Even more weathering was done with dust scraped from artist's pastel chalks. I used brown on the undersides to replicate dried mud stains and dirt. I used black for gunsmoke streaks and exhaust stains. The poor quality fuel used by late-war 109s didn't burn cleanly, so I made the stains really dirty and sooty. Another shot of Dullcote protects the pastel dust.

Itsy bitsies. After unmasking the canopy, I draped the left shoulder harness out of the cockpit. Next I added the underside details: landing gear FUG 16 antenna mast, pitot, aileron mass balances, and the IFF whip antenna. I replaced the kit tires with True Details resin ones and super glued 2.2-pound monofilament brake lines to the struts, **18**. I curled a bit of fly-tying wire for the end of the FUG 16 antenna and used a nylon paintbrush bristle for the IFF whip.

More 2.2-pound monofilament was used for the radio antenna topside. This was secured into predrilled holes with tiny drops of super glue. Waving a heated nail underneath the monofilament made it taut. I added tiny drops of white glue to simulate insulators. The loop antenna on the fuselage spine is a strip of coated bond paper that was curled around a dowel, painted brown-violet, and attached with white glue, **19**.

The canopy was attached with white glue and the joint overcoated with Dullcote to avoid the glossy glue look. Fly-tying wire was used for the canopy retaining cable.

Just when you think you're finished, something goes wrong. I had left off the propeller, figuring to add it last. Oh no! The spinner backplate didn't match the fuselage; the right "cheek" area of the nose was out-of-round. I ended up having to reshape the nose with emery boards and sandpaper, **20**. I then masked, repainted, and re-weathered the area. The addition of the prop finished the job.

And there you have it: a Messerschmitt that looks the part of a war-weary Luftwaffe veteran during the Third Reich's final hours.

Jim's updated Revell-Monogram kit, with its aftermarket parts and extra attention, is a striking representation of one of the Third Reich's final fighters.

LATE-WAR LUFTWAFFE COLORS

By the time production deliveries of the Bf-109K-4 commenced in mid-October 1944, Luftwaffe camouflage options had shifted to the so-called "late war" colors of 81/82/76 (brown-violet and green upper surfaces over light blue undersurfaces). However, an interim scheme of 75 (gray-violet)/81/76 persisted, probably because transportation difficulties and stringent austerity measures forced the use of existing paint stocks.

A third color, 83 (brown-green), introduced in August 1944, remains enigmatic as to extent or real intent of use, but paint samples gathered from surviving airframes shows it was sometimes used in conjunction with either 81, 82, or 75 for upper and side surfaces. Detecting 83 in black-and-white photos, however, is almost impossible.

Argument still rages over whether a fourth color, a distinct green-blue, was used to replace the pale gray of the late-war 76. The green-blue did blend extremely well with 81, 82, and 83, and some photos of Bf-109Ks show a distinctly warmer, deeper shade than 76.

No documents of camouflage directives have been found listing or describing this unidentified color – but then no camouflage directives have been located for the Bf-109K either. Still, late-production aircraft had some distinct patterns, suggesting that some standard was followed. While initially the K pattern mirrored that found on the G, eventually some changes were made.

The most obvious was the camouflage on the sides of the fuselage that was extended down to the bottom line of the Balkankreuze. On the port wing, the area of the darker color was increased. These measures increased the effectiveness of the camouflage while the airplane was on the ground.

A three-color pattern appeared on the horizontal tail and wing of some aircraft. The third color appears to have been 76 or 02 (greenish gray).

Hybrid schemes also occurred because subcontractors painted their subassemblies. In particular, completed tail units were delivered fully camouflaged which sometimes resulted in distinctly mismatched schemes. Field replacement of components such as elevators, rudders, and ailerons produced further mismatching of both pattern and colors. The stenciled fin and rudder pattern seen on some late model Bf-109Gs and K-4s is easy to detect in some photos.

– Ken Merrick

Nocturnal
STUKA

Converting Hasegawa's 1/48 scale Ju87D-5 into a D-8 BY ALFONSO MARTÍNEZ BERLANA

I enjoy building the latest generation of 1/48 scale kits, but sometimes building them right from the box doesn't provide much challenge. Lately, I've been searching for minor conversions that will set my models apart from the out-of-the-box crowd.

That was the case with my late-model Stuka. Using the excellent Hasegawa Ju 87D-5 (kit No. JT53), I transformed it easily into one of Germany's specially modified night-attack aircraft. Along the way, I added aftermarket interior parts to make the model even more special.

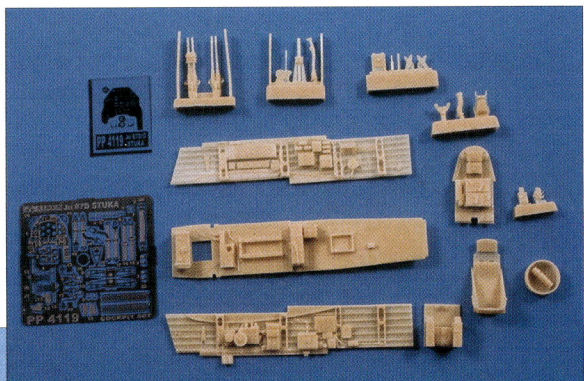

To improve the cockpit interior, I incorporated the Aires resin and photoetched detail set. The instrument panel is photo-etched with a printed acetate gauge set, which produces realistic instruments in this scale.

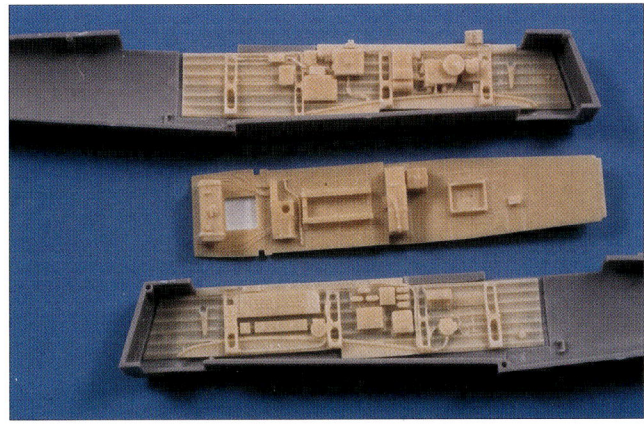

The resin cockpit tub is flanked by the sidewall detail parts glued to the Hasegawa fuselage halves.

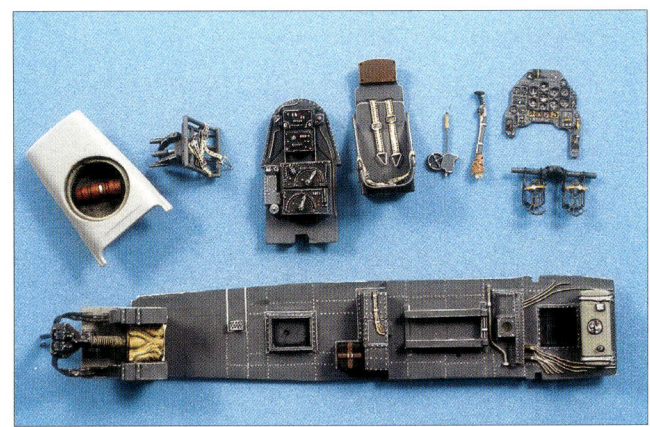

I built each interior unit as a subassembly and painted each separately. I used Life Color paints (available in Europe), but other brands will do. The base color of Luftwaffe interiors is schwarzgrau (black gray), which I accented with a black wash and light dry-brushing.

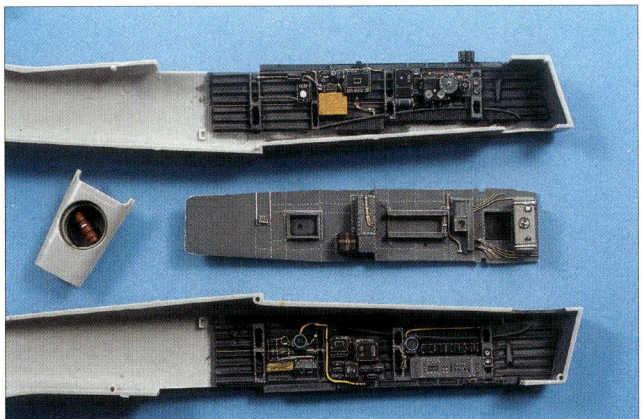

Here, all the details molded into the resin side panels have been painted. Once the floor and instrument panel are installed, I am ready to close the fuselage.

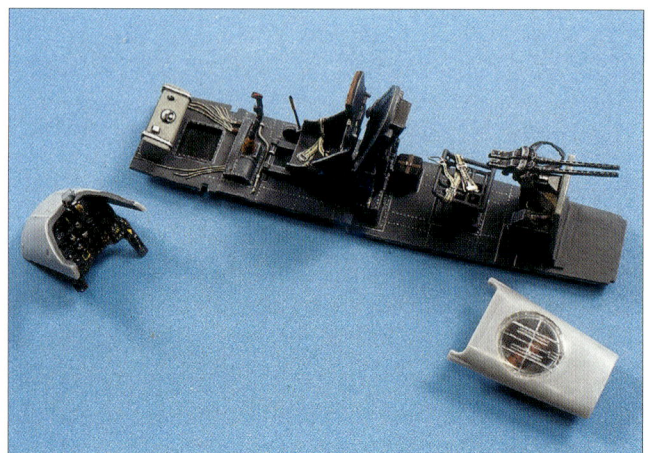

The seats, control stick, and rear guns are attached to the floor, and the instrument panel is in its coaming. The clear hatch houses a direction-finding antenna.

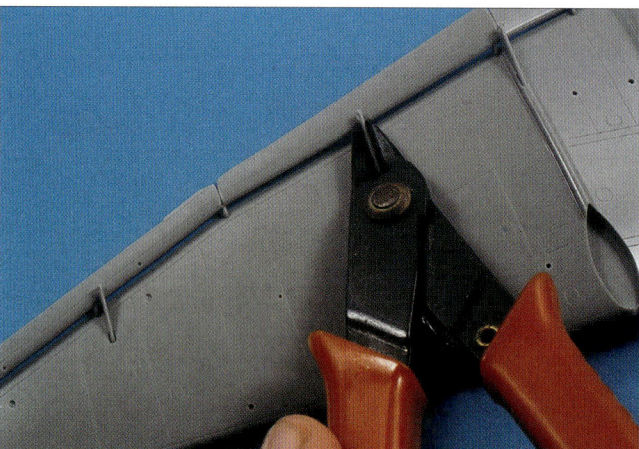

I removed the flap actuators from the bottom surface of the wings and replaced them with new ones made from stretched sprue.

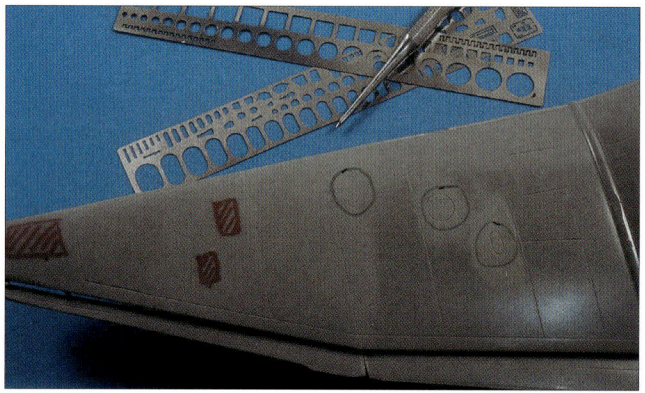

Hasegawa's wing had panel lines and access hatches that were typical of a Ju 87B. After studying my references, I filled and sanded smooth some of the features, then scribed in others that were missing.

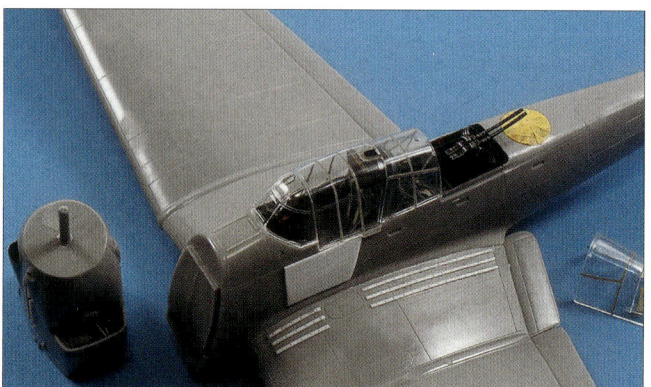

For better detail, I applied cockpit armor panels and wingwalk no-slip strips made from sheet styrene.

NIGHT ATTACK STUKAS

At the end of 1943, the Luftwaffe created several night-attack groups (Nachtschlachtgruppen) whose mission was an answer in-kind to the annoying nocturnal incursions of Soviet Polikarpov Po-2s on the Eastern Front. The initial aircraft assigned to the groups were Arado 66 and Gotha 145 biplanes, but later, the more effective Stuka was brought into the units. By that time, daylight ground-attack missions were being handled by fast Fw 190Fs and Gs, but the Stukas' slow speed was not a detriment for night attack sorties.

About 300 Ju 87D Stukas were modified for night attack. Flame-dampeners were attached to the exhaust pipes, flash hiders were added on the cannon muzzles, and since night tactics of low-level straight line bombing were the norm, the dive brakes were eliminated. Many of these machines wore strange camouflage, perhaps to hide the distinct outline against moonlit clouds.

The attacks were usually made during moonlit nights or at dusk. One or two Stukas would mark the target areas with flares, and the remaining aircraft would attack with bombs and cannons.

Modified Ju 87D-3s were redesignated Ju 87D-7s, and extended-span D-5s were redesignated D-8s. Nachtschlachtgruppen 1, 2, 4, 8, 9, and 10 used the night Stukas on the Eastern and Western Fronts and in Italy. Several groups used them until the end of the war, when advancing Allies found them on abandoned airfields.

– *Alfonso Martínez Berlana*

REFERENCES

Ju 87 Stuka in Action
Brian Filley, Squadron/ Signal Publications, Carrollton, Texas, 1986

Aero Detail No. 11, Ju 87 Stuka Dai Nippon Kaiga, Tokyo, Japan, 1994

Right: Alfonso made a stencil and painted a yellow 3 on the nose to represent a night-attack Stuka that saw action in Czechoslovakia late in World War II.

The gap between the kit spinner and the front of the nose was too large. The simplest solution was to extend the nose with a disc of .010" sheet styrene.

The first bit of painting was to apply the interior canopy brace color of black gray, sprayed over the masked canopy.

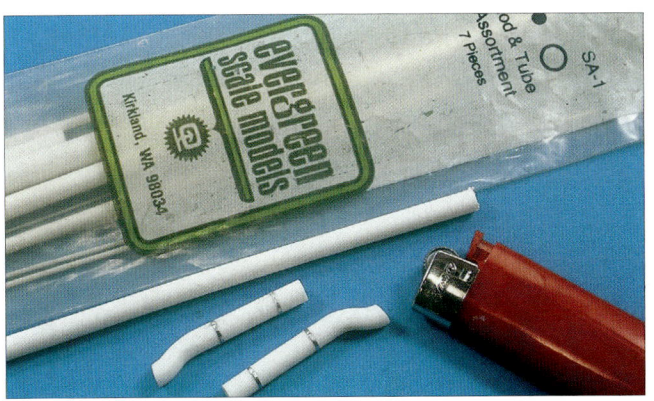

I formed the exhaust flame dampeners by heating Evergreen styrene tube with a butane lighter and bending it to shape.

The rippled camouflage was applied with a fine-tipped airbrush, thinned acrylic paints, and low air pressure. With an enamel base, you can remove acrylic paint errors with alcohol or water without marring the base coat.

Detailing Hasegawa's
ME 163B

Bring out the beauty of this 1/32 scale Komet   BY WILLIAM A. STEIDL   PHOTOS BY THE AUTHOR

The rocket-powered Komet packed a punch – and William Steidl's model is packed with detail, most of it scratchbuilt.

The Messerschmitt 163B Komet made its debut in 1941 as the world's first (and only) rocket-powered fighter. Though hard to handle and often dangerous to its pilots, it remains one of the most intriguing aircraft ever conceived – and a popular model subject. I build in 1/32 scale, and saw an opportunity to improve the Hasegawa kit by adding plenty of realistic detail – most of it scratchbuilt.

As it is with most 1/32 scale projects, selecting the particular markings for my Komet was a challenge. Eventually,

my resources on the Me 163B led me to build W.Nr191659 – it required the least amount of alterations, had the nicest color scheme, and Super Scale International offered decals (at the time, anyway).

To keep the project organized and give myself some sense of accomplishment, I divided the model into subassemblies: the cockpit, the landing gear, the engine, and the fuselage.

Constructing the cockpit. First, I collected all of the kit's cockpit parts and

dry-fitted them. This gave me ideas about the relative space between items and where there would be room for additional detail. Then I pulled out my reference material and decided which parts were going to stay and which parts weren't.

I removed the cockpit tub's raised detail, which was mostly the strapping for the T-Stoff fuel tanks (located on either side of the tub) and the radio panel. I then added my own scratchbuilt strapping, **1**, and radio panel, **2**. The armor plate from the kit was just fine, plus it ensured a good fit within the fuselage. From there I moved on to the oxygen regulator, which I also decided to rebuild, **3**.

The trim control box and hand wheel were rebuilt using styrene, a plastic washer, and stretched sprue, **4**. I modified the kit's control stick by running wire along the length of the shaft (secured with lead foil), using foil to make a machine-gun trigger at the top of the handle.

The Komet's gun sight is conspicuous, so I wanted mine to look its best. I started out with 3/32" sheet styrene and cut my two basic blocks using the kit part as reference for dimensions. I then drilled a 1/16" hole, painted the edges silver, filled it with epoxy, and topped it off with a Waldron instrument bezel. The second block was then attached to the first, keeping the left sides flush. The kit glass was used for the reflector plate and some old film for the tinted glass plate. I finished the gun sight by cutting and attaching lead foil as a frame for the glass, **5**.

I used the kit's main instrument panel as a guide for constructing a new and improved panel out of styrene. First, I marked where the gauges would go, punched the holes, and painted the panel. After it dried, I punched and placed the appropriate Waldron dial face in each of the holes, and then glued from the back – a very clean method.

I finished the panel by gluing Waldron instrument bezels around each of the dial faces, and then dry-brushing them. I used stretched sprue for switches and black wire for knobs. I like to pre-drill all of my switch locations (I prefer a No. 80 thumb drill), insert the sprue, and then glue from behind. Old computer chips can be recycled into great throttle levers – just snip a "leg" off the chip and attach 1/32"-diameter punched disks (from .015" sheet styrene) to either side, **6**.

To improve the cockpit seat, I rounded off its corners, added seat belts and buckles from Waldron, and formed a seat cushion from clay, **7**. I finished off the cockpit by adding some ribbing to the sides of the fuselage using lead foil, some fuel lines using lead wire, and two more levers for the canopy release and tow cable release.

Here's a little trick: Glue the cockpit tub to one side of the fuselage and the

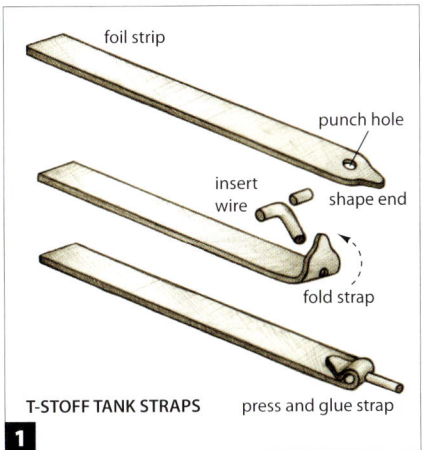

T-STOFF TANK STRAPS

1

William made strapping for the T-Stoff tanks from lead foil strips and 24-gauge wire. After the straps were in place and the cockpit was painted, he removed paint from the straps with dirty black thinner.

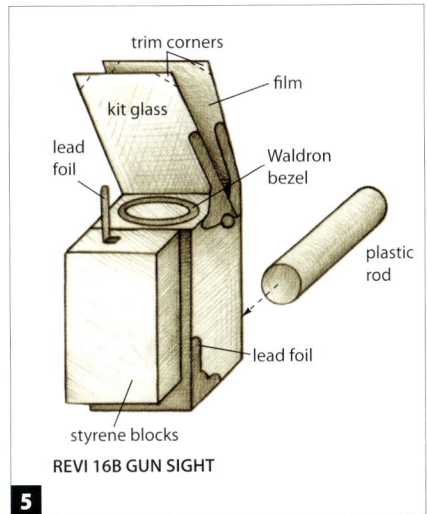

OXYGEN REGULATOR

3

William built the replacement oxygen regulator from styrene and wrapped wire, securing the ends of the wire with a thin strip of foil and making a connection-like assembly.

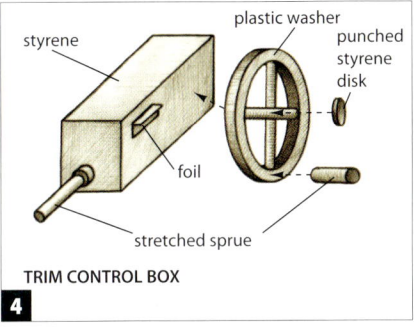

REVI 16B GUN SIGHT

5

The Revi 16B gun sight was built using sheet styrene for the basic block, drilling out the lens hole, filling it with epoxy, and adding a bezel.

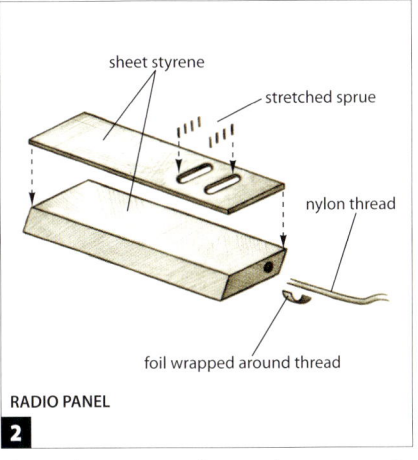

RADIO PANEL

2

The replacement radio panel was topped with a 1/64" sheet containing cutouts for switches, made by punching out 1/32"-diameter holes at either end of the slot and cutting out the pieces between the holes.

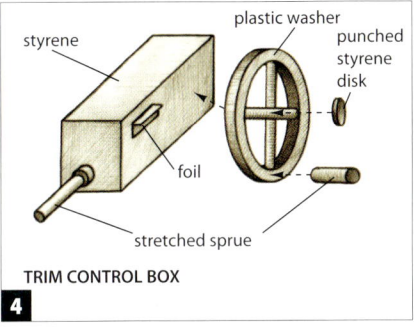

TRIM CONTROL BOX

4

For this trim control box, William used materials that gave him the look he wanted. Lead foil, still found on some wine bottles, proved to be a handy material for this project.

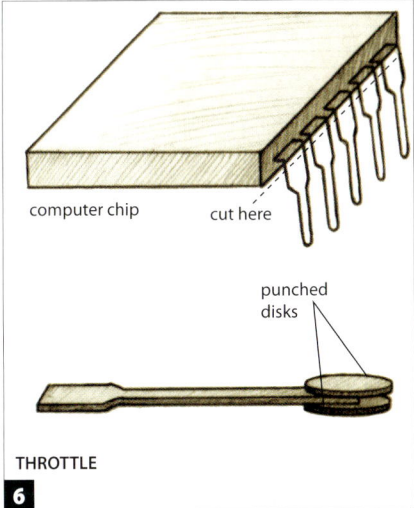

THROTTLE

6

Throttles are easy to make – just attach two 1/32" diameter disks of .015" styrene sheet to either side of a computer chip "leg."

7

William filled every available space in the cockpit with carefully planned details. Skillful dry-brushing makes the details stand out realistically.

8

The kit includes only a basic wheel dolly. Since you can't just pull one of these out of your spares box, research is essential to getting it right.

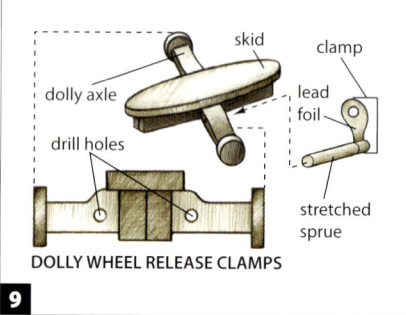

DOLLY WHEEL RELEASE CLAMPS

skid
clamp
dolly axle
lead foil
drill holes
stretched sprue

9

The dolly clamps were made by folding a rectangle of lead foil in half, cutting out a teardrop shape, then inserting a 1/16"-diameter rod through the unfolded shape.

10

In the kit part, the tail wheel and the fork were molded as one. William's replacement makes a dramatic difference.

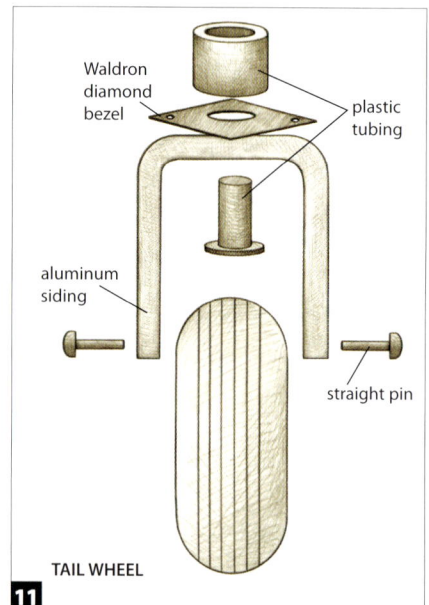

Waldron diamond bezel
plastic tubing
aluminum siding
straight pin

TAIL WHEEL

11

After a fork made of plastic sheet proved too weak to support the weight of the plane, William rebuilt the tail-wheel fork from leftover parts of aluminum siding trim.

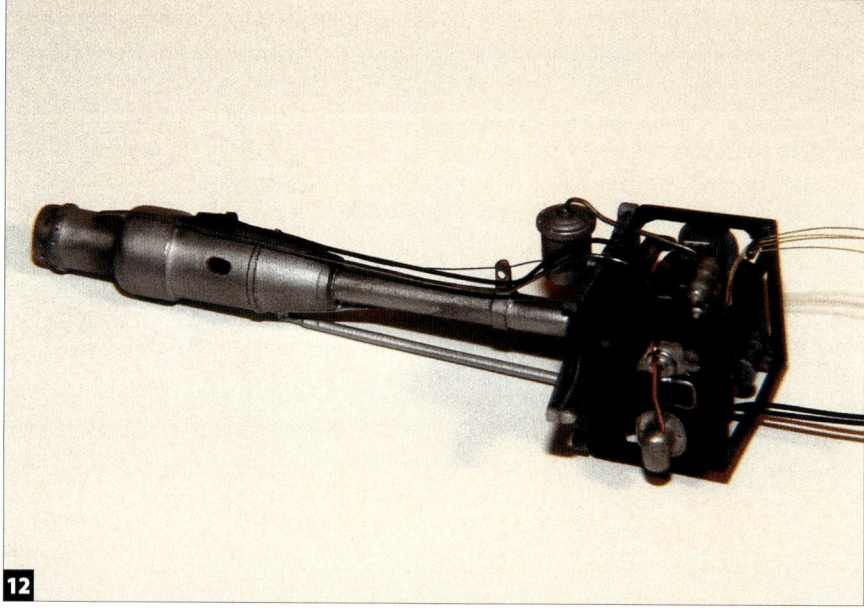

12

The engine was finished according to references, including all the wiring. You can't see it on the finished model, but it's in there.

Right: The Komet, equipped with a mismatch of slow guns and a rocket-powered engine, was a technological dead end but an impressive sight over German skies.

instrument panel to the other side. They won't interfere with one another when they are glued, and it's much easier to glue the instrument panel to the fuselage.

The canopy was lined with lead foil on both sides, and an air intake hole was added to the front. Not all of the Me 163s have this air intake at the front of the canopy, so you'll need to check references for the particular plane you're building.

Landing gear. The landing gear was modified by adding a dolly release mechanism and a reinforced skid plate, **8**. The kit has no detail on how the wheel dolly gets attached to the skid in real life – the dolly is missing two clamps from its detail and the skid lacks a release mechanism. The clamps attach to protruding rods on either side of the skid. The idea was for the rods to retract after takeoff, releasing the dolly. The clamps, made from stretched sprue and lead foil, **9**, were attached to the dolly axle by drilling out holes on either side.

Since the kit's tail wheel was molded into the fork, I completely rebuilt the whole assembly, **10**. I modeled it first in thick paper following my references as a guide, then transferred the template over to aluminum and drilled out all of the holes. Next, I bent the aluminum into its proper shape. It was strong enough, but didn't take well to gluing, so I drilled a hole through the center of the fork and ran a capped plastic rod from the bottom, through the hole, and into a plastic tube above, **11**. The arm connecting the tail wheel to the plane was cut from plastic stock using the kit-supplied part as a guide for thickness.

Engine. The engine was built from the kit parts, but then I went on a wiring frenzy and added all the engine lines I could, **12**. The sad part is that no one will ever see all the work done to the engine because I elected to glue the front and rear sections of the fuselage together!

Finishing the fuselage. I sanded off all the raised detail on the fuselage and the wings and re-scribed the entire plane. This took a while, and I highly suggest to anyone who is thinking about doing such a job to perform this task while the wings and elevators are not yet joined. It just makes the job so much easier because everything can be laid down flat and scribed. I created a set of templates for scribing the irregularly shaped hatches.

Using Testor Model Master enamel paints, I custom blended each of the colors, matching the paint chips supplied in Merrick and Hitchcock's book (see references). The Luftwaffe's camouflage scheme called for a splinter pattern of RLM 81/82 (brown-violet/dark green) on all upper surfaces, and RLM 76 (light blue) for the tail and under surfaces. I switched from a single-action to a dual-action airbrush to apply a mottling pattern of RLM 81/82 to the tail. I stencilled and sprayed all the markings except for the two 15s, which were decals. The production numbers and fuel type letters were dry transfers.

An ordinary lead pencil was used to darken the panel lines; following up with pastels provided the shading. The entire model then got a coat of Testor Model Master clear flat, which provided the proper sheen.

REAL-LIFE REFERENCE

Out of the hundreds of Komets produced for the Luftwaffe, about a dozen remain on display in museums around the world. In December 1999, the United States Air Force Museum (Dayton, Ohio) acquired an Me 163B Komet from the National Aviation Museum of Canada. Other Komets are owned by the Aerospace Museum (Cosford, England), the Australian War Memorial (Canberra, New South Wales), the Canada Aviation Museum (Ottawa, Ontario), the Deutsches Museum (München, Germany), the Eighth Air Force Heritage Museum (Macon, Georgia, on loan from the National Air and Space Museum), the Imperial War Museum (Duxford, England), the Luftwaffenmuseum (Berlin-Gatow, Germany), the Museum of Flight (East Fortune, England), and the National Museum of Science and Industry (London, England).
— *Elizabeth Lamb*

REFERENCES

The Official Monogram Painting Guide to German Aircraft 1935-1945 Kenneth A. Merrick and Thomas H. Hitchcock, Monogram Aviation Publications, Boylston, Massachusetts, 1980

German Aircraft Interiors 1935-1945 Vol. 1 Kenneth A. Merrick, Monogram Aviation Publications, Sturbridge, Massachusetts, 1996

Aero Detail 10: Messerschmitt Me 163 and Heinkel He 162 Nohara and Shiwaku, Dai-Nippon Kaiga Publishing, Japan, 1994

Messerschmitt Me 163 "Komet" Vol. 2 M. Emmerling and J. Dressel, Schiffer Publishing, West Chester, Pennsylvania, 1992

Cockpit Donald Nijboer, Howell Press, Charlottesville, Virginia, 1998

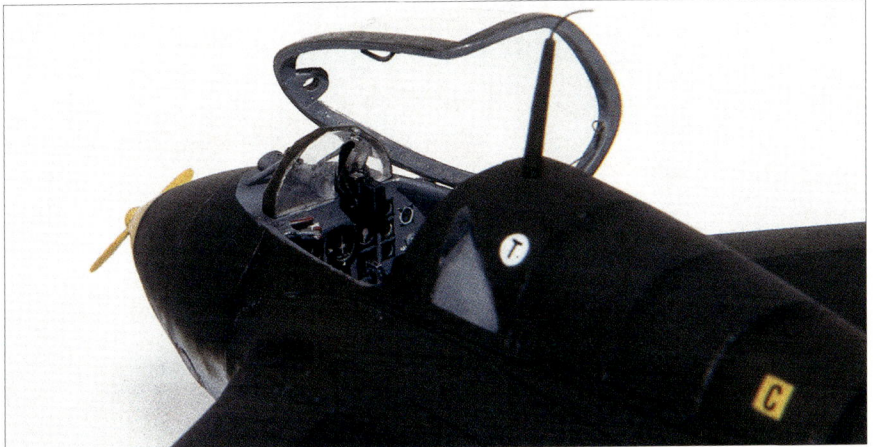

Superdetailing Revell's classic 1/32 scale NIGHTFIGHTER

When outfitted with radar, supplementary fuel tanks, and upward-firing cannons, the often-criticized Bf110 "Zerstörer" became a potent weapon   BY JIM FULLINGIM

evell's old big-scale Messerschmitt Bf 110 wasn't bad for its day. It has been reissued in several versions, but I wanted to make the late Bf 110G-4c/R8 with its improved FuG 220b Lichtenstein SN-2 radar antennas.

Radar and weapons fit varied greatly among late-war nightfighters. The R8 version (R stood for Rustsatze or auxiliary apparatus) included upward-firing Schräge Musik (jazz music) cannons in the rear cockpit. The antennas, weapons, and many other details differed from the kit parts, so I had plenty of work to do.

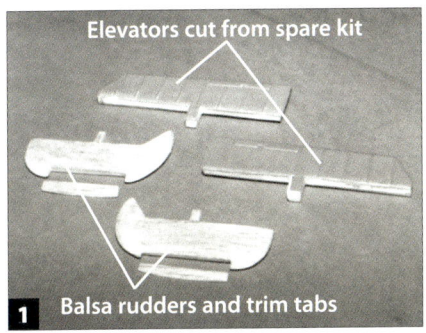

Elevators cut from spare kit

Balsa rudders and trim tabs

1

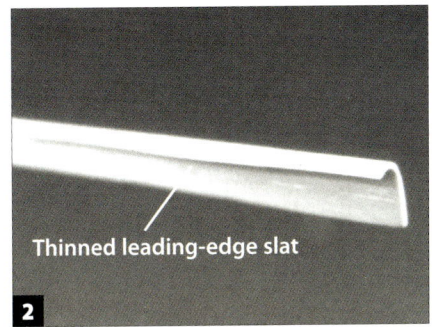

Thinned leading-edge slat

2

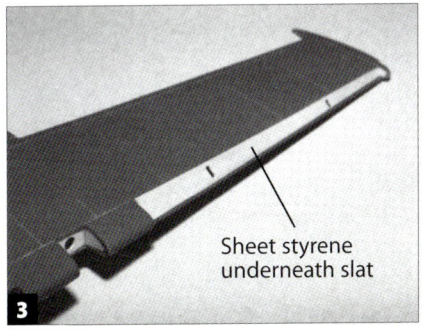

Sheet styrene underneath slat

3

Preparation. After sanding away all the raised surface detail and penciling locations for new lines, I scribed recessed detail with a Bare-Metal Foil scriber guided by a metal ruler, bent copper sheets, and an artist's erasing shield.

I cut away all the control surfaces so I could position them. This operation required two kits. Cut the surfaces from the spare kit with a little extra room at the cut edges. You can comfortably shape the leading edges so they appear realistic, **1**. Keep the rest of the spare kit as it may come in handy.

Cut the surfaces from the main kit, but don't remove too much. Dress the opened

1/32 SCALE

4

edges with sheet styrene and filler.

I also cut away the leading-edge slats, filed and sanded them to scale thickness, **2**, and covered the area underneath the slats with sheet styrene, **3**. Revell's rudders

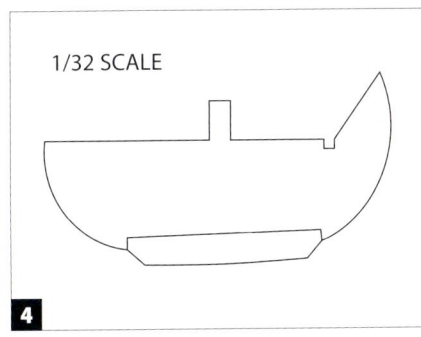

Balsa rudder (tab added later)

After

Before

Remove kit rudder

5

were not accurate for the G-4 version, so I made new ones from balsa, **4** and **5**, then sealed the wood with Pactra balsa sanding sealer.

Nose modifications. I established the outlines for new floorboards and bulkheads for the cockpit and nose gun platform by bending electronic solder to fit inside the fuselage. I used these contour gauges to pencil the shapes on sheet styrene, then cut and filed them to shape.

I "fattened" Revell's pair of Mk. 108 nose cannons by adding bottoms cut from the spare kit's cannons, **6**. (See, those extra parts do come in handy!) The ends of each cannon barrel were replaced with

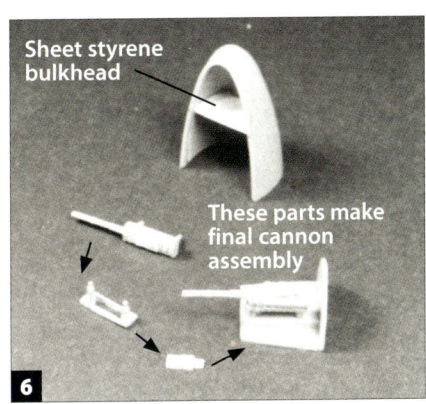

Sheet styrene bulkhead

These parts make final cannon assembly

6

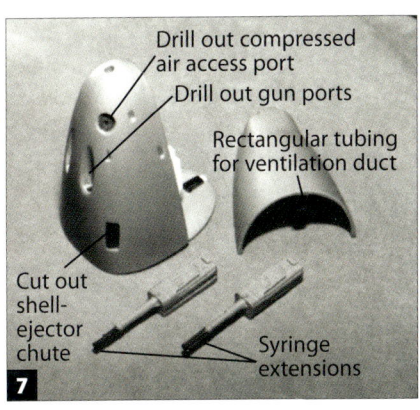

Drill out compressed air access port

Drill out gun ports

Rectangular tubing for ventilation duct

Cut out shell-ejector chute

Syringe extensions

7

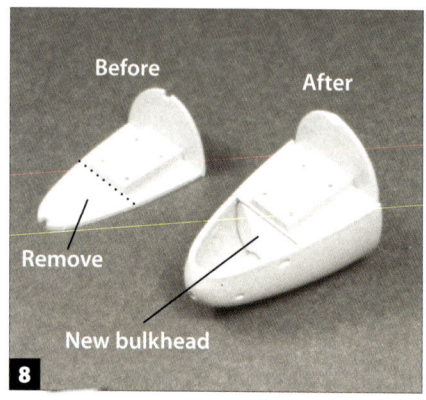

Before **After**

Remove

New bulkhead

8

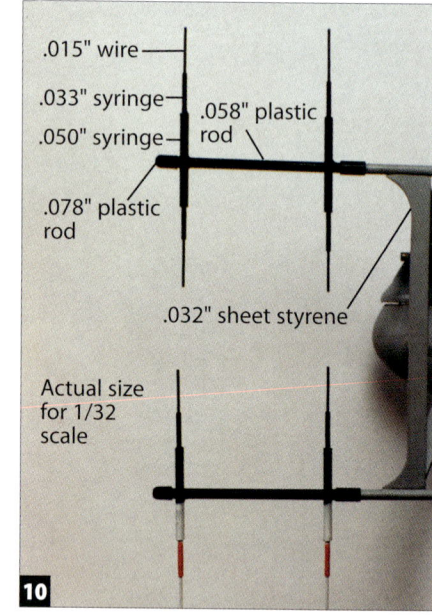

Eduard photoetched parts

9

Compressed-air bottles made from
dowels, plastic, and brass fittings

pieces of medical syringes (.056" and .069" diameters), **7**.

I raised the forward cannon deck with sheet styrene so the guns would project from the ports in the nose. The cannons also were staggered horizontally to allow clearance from the ammunition feed chutes.

I drilled a hole in the center of the nose, squared it with files, then attached square brass tubing for the pilot's fresh-air intake.

A bulkhead, **8**, under the forward end of the gun deck forms the rear of a compartment that holds compressed-air bottles made from dowel, sprue, and MSC wire, **9**. I used square brass tubing again for the shell-ejector chutes.

I scribed circular covers for the access valve to the compressed-air-bottle recharging station on the left side of the nose and below the left side of the pilot's windshield. I also inserted two sections of .062" aluminum tubing for the lower MG 151/20 nose cannons after drilling out holes in the kit's cover.

The kit's radar antennas and mounts weren't correct for this aircraft, so I enlarged my photo references on a copier to 1/32 scale, then reconstructed this antenna array with medical syringes, steel wire, and styrene, **10**.

.015" wire

.033" syringe
.050" syringe
.058" plastic rod

.078" plastic rod

.032" sheet styrene

Actual size for 1/32 scale

10

Cockpit. I kept the kit's sidewalls, but otherwise rebuilt the cockpit, **11**, using lots of homemade photoetched parts. I drew each item eight times the normal size, then ordered scale-size negatives for producing the brass parts. I also used several Eduard photoetched parts. Instrument faces were punched from Waldron's 1/32 scale German instruments set.

The pilot's seat and other parts came from Verlinden Products' Bf 109E update set (No. 741). The roll bar and frame supports were made from styrene rod. I fashioned the pilot's throttle console from balsa, with metal rods for throttle levers, **12**. All balsa panels were sanded and sealed with Pactra balsa filler coat, which left the surfaces plastic-smooth. The knobs were cut from styrene rod.

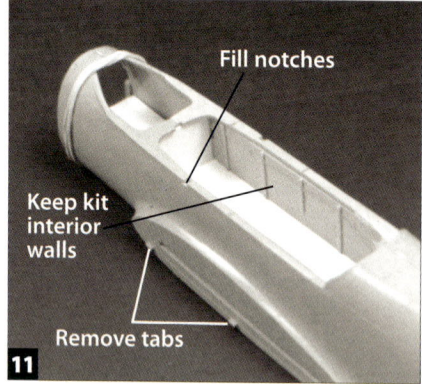

Fill notches

Keep kit interior walls

Remove tabs

11

Each radio and radar box was fashioned from balsa and my photoetched parts, **13**. I installed a pair of blue light-emitting diodes (LEDs) for the radar screens, **14**. They eventually were hooked up with other lights and a power supply; more on that later.

Elevator trim-tab and wheel

Throttle console

12

I made ammo boxes on each side of the radar operator's cockpit from balsa with homemade photoetched tops and installed nearby avionics details from Eduard, **15**.

The wing spar in the cockpit is aluminum tubing and bent .100" sheet styrene. I made the radar operator's seat (on top of the spar) from brass tubing, styrene, and interlaced strips of masking tape. For scale effect, I placed a flare pistol from the Tamiya 1/25 German infantry weapons set on top of the seat. The seat cushion (leaning against an ammo box on the floor) and a similar cushion for the pilot were sculpted from Milliput two-part epoxy putty.

One of the most difficult areas to detail was the rear-cockpit installation of the upward-firing 20mm cannons, **16**. I used gun parts from Hasegawa's 1/32 scale Fw 190A. I also utilized medical syringes, LEDs (just for the shape), photoetched parts, and ¼" wooden screw-hole plugs (perfect for the ammunition drums; just file off the bottom halves). This unit then was installed in a 70-degree firing angle, **17**.

The rear-firing twin machine guns were made with MG34s from Tamiya's 1/35 scale German infantry weapons set. They were clamped together with styrene tube and feature Eduard details. The assembly was mounted on a dowel.

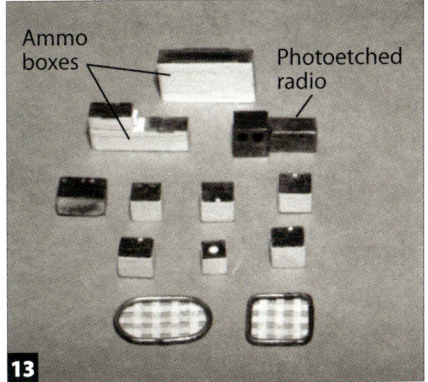

13 Ammo boxes · Photoetched radio

14 Blue LEDs for FuG 220 SN2 radar

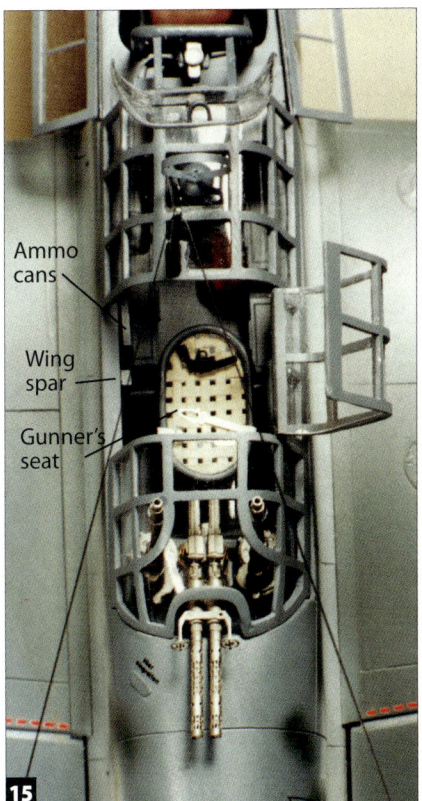

15 Ammo cans · Wing spar · Gunner's seat

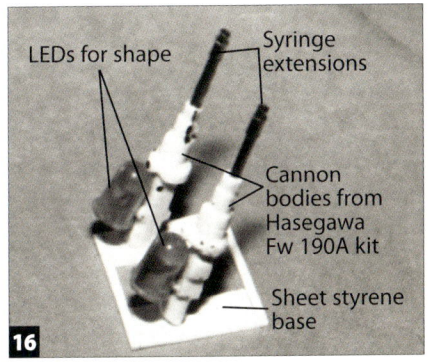

16 LEDs for shape · Syringe extensions · Cannon bodies from Hasegawa Fw 190A kit · Sheet styrene base

17 Ammo canister from wood screw-hole plug, plastic, and photoetched parts · Ammo container made from dragster fuel tank

Canopy. The kit's seven-piece canopy was thick and distorted, so I made a new one. First, I made masters by filling the kit canopy sections with Sculpey III modeling clay, **18**. After removing the clay parts and refining their shapes, I baked them in the kitchen oven until hard.

Next, I vacuum-formed clear copies of each section with my old Mattel Vac-U-Form, **19**. I trimmed the new canopies to fit inside the kit parts and drilled holes for the Schräge Musik guns.

18 Sculpey masters made from kit canopy parts

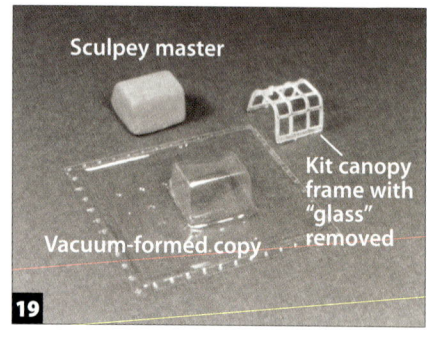

Sculpey master

Kit canopy frame with "glass" removed

Vacuum-formed copy

19

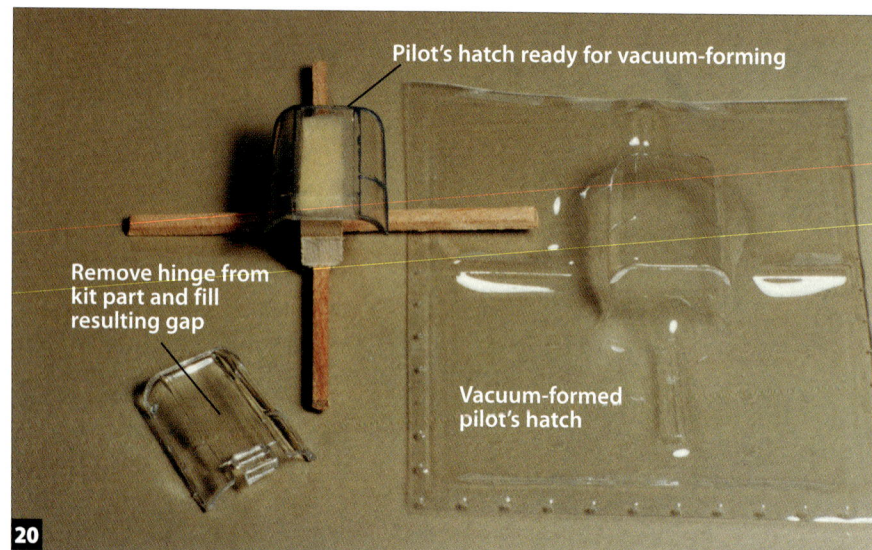

Pilot's hatch ready for vacuum-forming

Remove hinge from kit part and fill resulting gap

Vacuum-formed pilot's hatch

20

The next stage was the most difficult. I used the kit canopy parts for the framing. I drilled and filed out all the clear areas, with the exception of the center windscreen and the pilot's hatch. These skeletal frames were carefully sanded to scale thickness.

I corrected the shape of the pilot's hatch by cutting away the hinge, adding a small section of clear styrene, then vacuum-forming a copy, **20**. I remade the pilot's right and left folding windows with clear and strip styrene, and replaced the left and right panels of the windscreen with clear sheet.

Revell incorrectly depicts the gunner's and radar operator's hatch hinges; they should be higher on the right side of the canopy, so I corrected them, too, **21**.

This canopy replacement system allows me to paint the frames without masking the clear panels. I added the clear panels and vacuum-formed sections with white glue after painting.

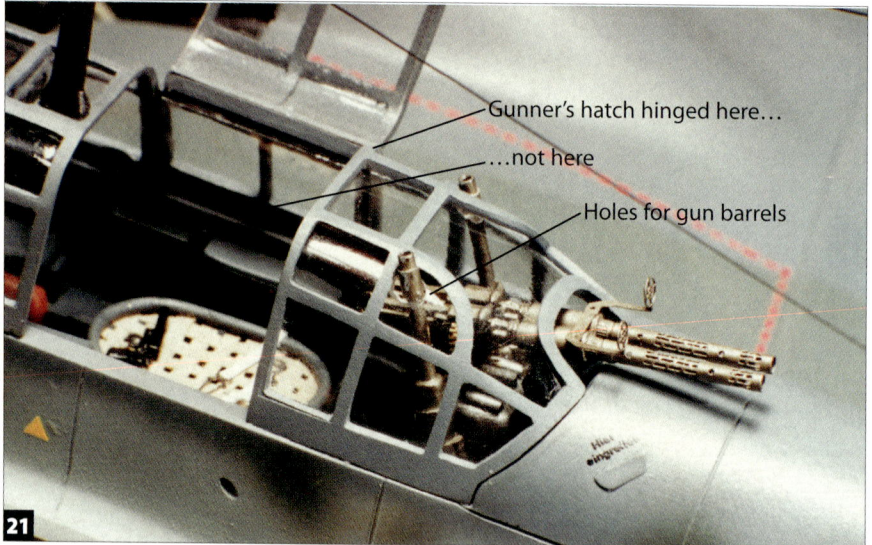

Gunner's hatch hinged here...

...not here

Holes for gun barrels

21

Engines and props. I wanted to expose one of the aircraft's engines but found the Revell engine unsuitable, **22**. Instead, I installed the engine from Verlinden's Bf 109E detail set and detailed it following my references, **23**. To show more detail at the rear of the engine, I cut open the cowling on the upper surface of the wing and installed the panel cut from the spare kit, **24**.

Revell engine

Verlinden engine

22

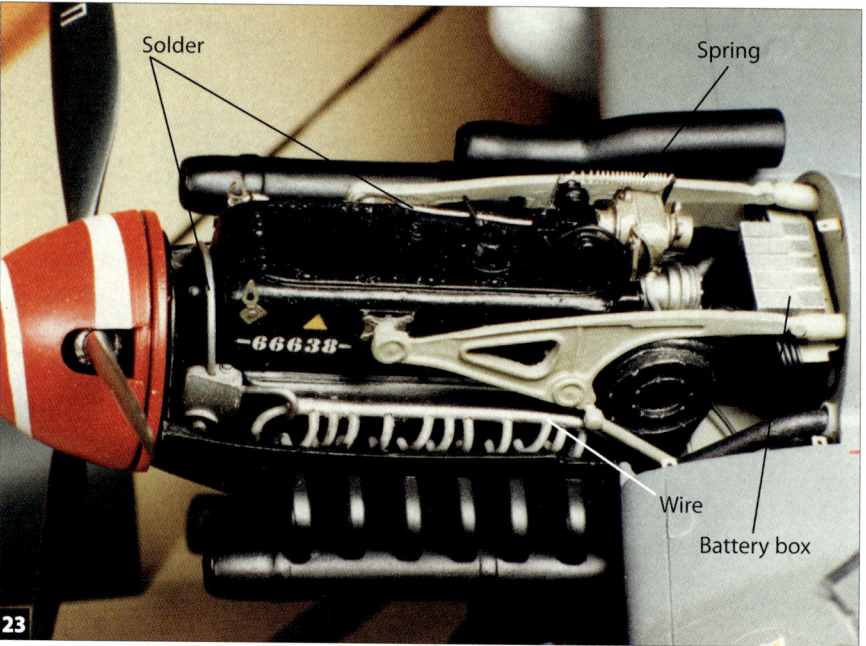

Solder

Spring

Wire

Battery box

23

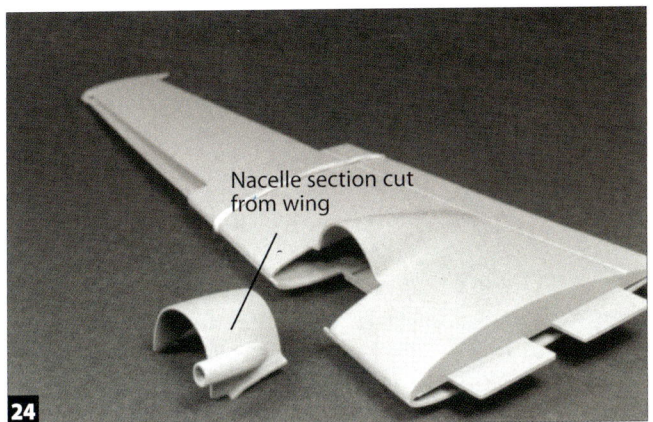

Nacelle section cut from wing

24

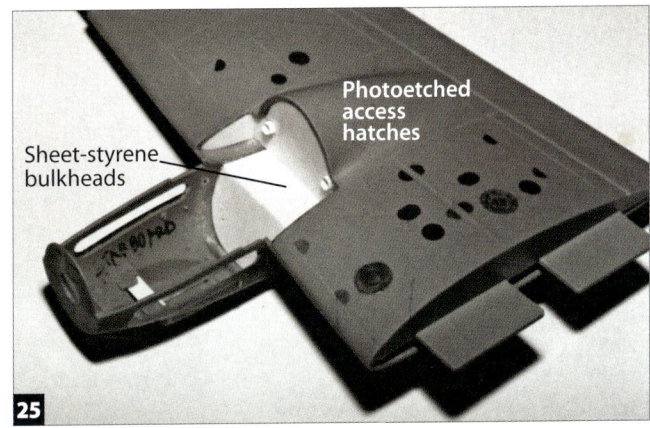

Photoetched access hatches

Sheet-styrene bulkheads

25

I fashioned a firewall and boxed in the landing-gear wells with sheet styrene, **25**. The large gap around the engine exhausts was corrected with homemade photoetched panels. I replaced the doors around the oil and glycol radiators with photoetched copies, **26**.

The Bf 100Gs had large Daimler-Benz engines with magneto placement that necessitated bulged fairings on top of the rear cowlings. I cut the bulges from a Revell Bf 109G kit – actually gun fairings on the nose. The four bulges were grafted onto the Bf 110G's nacelles and sanded smooth, **27**.

The propellers and spinners also were incorrect, so I replaced them with spares, **28**. The props came from a pair of Hasegawa Fw 190As and the spinners from two Revell Bf 109Gs.

Photoetched oil-cooler door (closed)

26

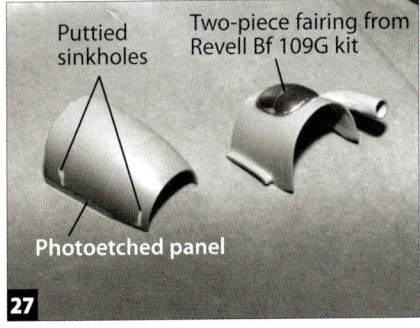

Puttied sinkholes

Two-piece fairing from Revell Bf 109G kit

Photoetched panel

27

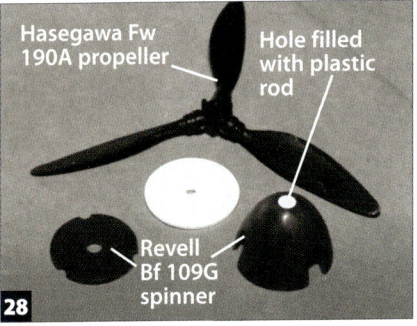

Hasegawa Fw 190A propeller

Hole filled with plastic rod

Revell Bf 109G spinner

28

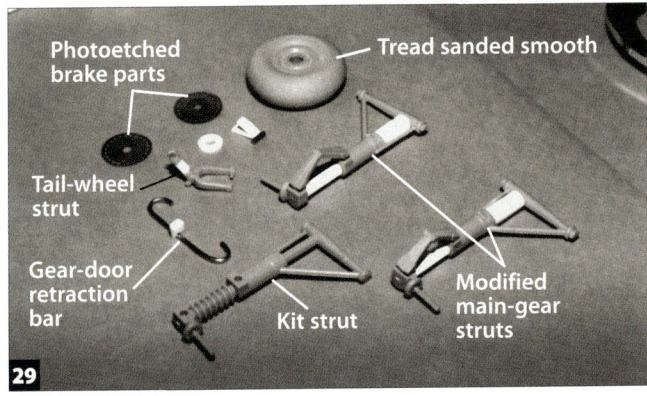

Photoetched brake parts

Tread sanded smooth

Tail-wheel strut

Gear-door retraction bar

Kit strut

Modified main-gear struts

29

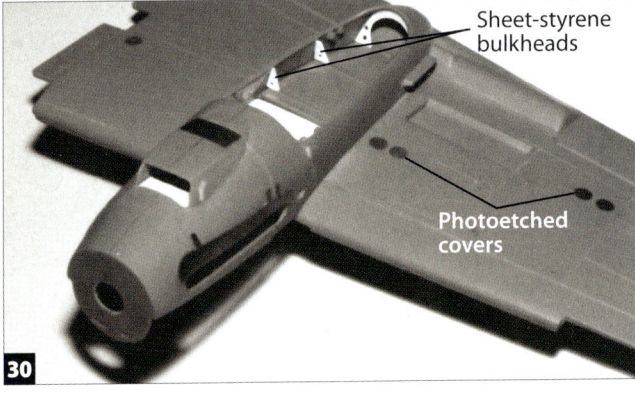

Sheet-styrene bulkheads

Photoetched covers

30

Landing-gear and wing details. Revell's landing gear was typical of earlier variants, so I reproduced the units visible in reference photos of the only surviving Bf 110G-4 in the Royal Air Force Museum in London.

I replaced the canvas-covered center sections with .125"- and .185"-diameter Plastruct tubing, **29**, and wrapped Bare-Metal Foil around the lower sections to depict the oleo struts.

I sanded the tread off the tires, then added homemade photoetched brake discs. Brake lines, made from .032" solder, were added later.

Interior braces made from plastic were added inside the landing-gear wells, **30**. I also made actuator bars that open and close the gear doors from brass tubing, scrap plastic, and Radio Shack insulated wire-wrap wire. The springs were made by stripping the insulation, then wrapping the wire 50 times around a stiff wire. When removed, the springs were trimmed to fit.

No underwing fuel tanks came with the kit, but I had an out-of-production Horizon Conversions vacuum-formed set that I detailed with scrap plastic, photo-etched parts, and MSC wire. (My wife said, "They look like dead katydids on their backs, with their feet stuck up in the air!" So much for realism.)

Many Bf 110G-4s had an external bomb rack that was left on many nightfighter versions. I deleted mine and scratchbuilt a lower plate with the shell-ejector chutes from .082" sheet styrene and square brass tubing. The pitot tube was made from a .038"-diameter medical syringe.

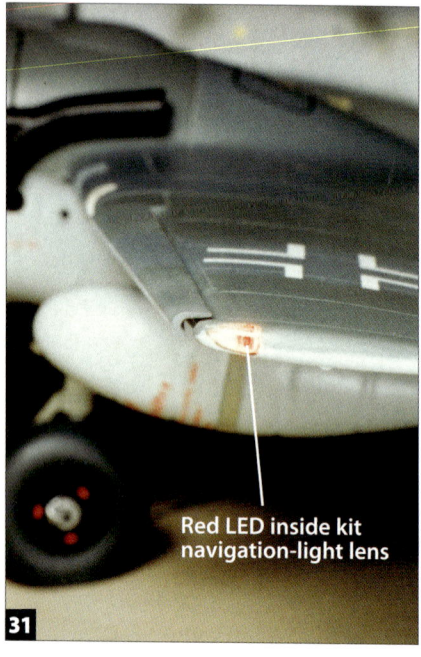

Red LED inside kit navigation-light lens

31

Electrical circuits exit through connectors in bottom of left tire

32

Electrical hook-ups. All the exterior lights work on this model. The clear tail-navigation light is a fiber-optic strand leading to a tiny Binkman mini-flashlight bulb under the cockpit floorboard.

The red (left) and green wing-tip navigation lights are LEDs filed to shape and inserted underneath the kit's clear lenses, **31**. Wires soldered to the LED terminals meet wires from the radar-screen LEDs in the left engine nacelle. They were soldered to female AMP avionics connectors buried in the bottom of the left main-gear tire, **32**. The wires thread through hollowed-out struts.

I installed another mini-flashlight bulb in the left wing leading edge for the landing light, **33**. The kit's light was incorrect, so I made a light box with sheet styrene covered with Bare-Metal Foil. The bulb projects from a hole drilled in the back of the box. The clear cover was produced by vacuum-forming clear plastic over a section of wing leading edge cut from the spare kit. This oversize lens was grafted into the leading edge with five-minute epoxy and masked for painting.

Wire leads from the landing light and the bulb for the fiber-optic tail-navigation light were soldered together and led through the strut into a separate set of connectors in the bottom of the tire. Separate circuits were needed as the bulbs required a different voltage than the LEDs.

Four corresponding pin connectors stick up about ⅛" from the model's base, **34**, and connect to a pair of twin AAA battery packs and a switch buried under the surface.

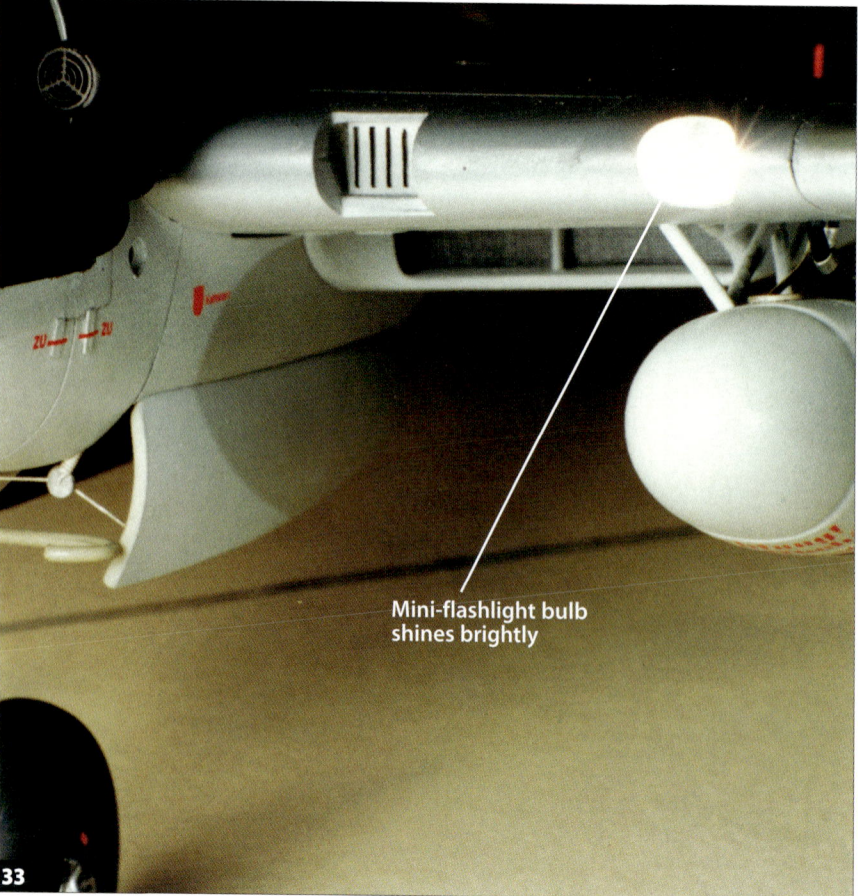

Mini-flashlight bulb shines brightly

33

Male connectors attached to base

34

Finish. The aircraft I modeled was a Bf 110G-4c/R8 of an unidentified unit – possibly N.J.G. based in Norway. The typical nightfighter camouflage of 1943-44 was a splinter pattern of RLM 74 gray-green (graugrun) and RLM 75 gray-violet (grauviolet) on upper surfaces, with fuselage sides, vertical stabilizers, and undersurfaces in RLM 76 light blue (lichtblau). A fine mottling of RLM 02 gray (grau), LM 70 black green (schwarzgrun), and RLM 75 was applied on the fuselage and vertical stabilizers. Propellers were painted RLM 70.

Nearly all of the markings were from a homemade set of dry transfers. The white swirls on the red spinners were stolen from SuperScale sheet No. 32-93. The dry transfers also included the lower engine oil coolers and lower wing radiator screens.

The artwork for these designs was drawn eight times larger than scale, reduced to 1/32 scale on lithograph negatives, then made into black, white, and red dry-transfer sheets. Some of the other colored stencils and warning flags were hand-painted.

After the markings were applied, I airbrushed several light coats of AeroMaster clear flat, then removed the mask from the landing light and installed the clear por-

tions of the canopy.

Red-and-white ground-crew warning stripes were masked and airbrushed on the lower rods of the radar antenna and the ventral FuG 25a antenna aft of the crew boarding ladder. The antenna wires were from stretched black sprue.

Whew! I worked on my nightfighter for the better part of two years, during which my wife and I moved twice! All that effort paid off, though, because the model won best large-scale propeller-driven aircraft at the IPMS Region VI Scalefest in May 1996.

REFERENCES

Bf 110G, Monogram Close-Up No. 18 George G. Hopp, Monogram Aviation Publications, Sturbridge, Massachusetts, 1986

The Defense of the Reich – Hitler's Nightfighter Planes and Pilots Werner Held and Holger Nauroth, Arco Publishing Co., New York, 1982

Famous Airplanes of the World, Messerschmitt Bf 110, No. 41 Burindo Co., Tokyo, 1993

Hitler's Luftwaffe Tony Wood and Bill Gunston, Crescent Books, New York

Instruments of Darkness Alfred Price, Peninsula Publishing, Los Altos, California, 1987

Messerschmitt Bf 110 at War Armand Van Ishoven, Ian Allan, Runnymede, England, 1985

The Messerschmitt Bf 110 Night Fighters No. 207 Profile Publications, London, 1972

Messerschmitt Bf 110 Over All Fronts, 1939-1945 Holger Nauroth and Werner Held, Schiffer Publishing, Atglen, Pennsylvania, 1991

The Messerschmitt Bf 110, Profile No. 23 Profile Publications Ltd., London, 1971

Messerschmitt Bf 110, Monografie Lotnicze No. 16 A.J. Press, Warsaw, 1994

Messerschmitt Bf 110 Zerstorer in Action Jerry L. Campbell, Squadron/Signal Publications, Carrollton, Texas, 1977

The Official Monogram Painting Guide to German Aircraft, 1935-1945 Kenneth A. Merrick and Thomas H. Hitchcock, Monogram Aviation Publications, Sturbridge, Massachusetts, 1980

Warplanes of the Luftwaffe David Donald, Aerospace Publishing, London, 1994

Detailing and painting Hasegawa's 1/48 scale "TONY"

Improving the sleek Ki-61 Hien   BY ALFONSO MARTINEZ BERLANA   PHOTOS BY AURELIO GIMENO RUIZ

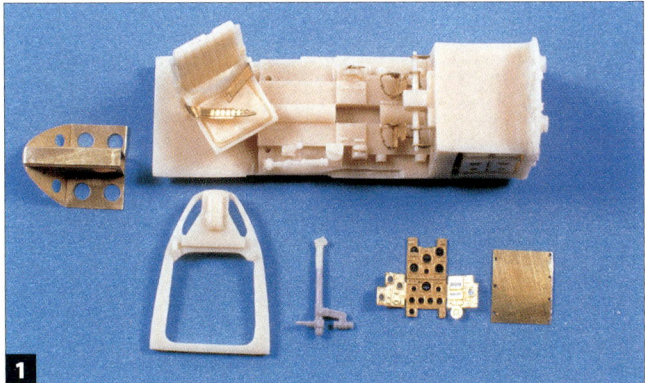
The Aires interior set includes resin and photoetched details.

Aires' set also includes the nose guns, bays, and covers.

The interior details were assembled, painted, and installed.

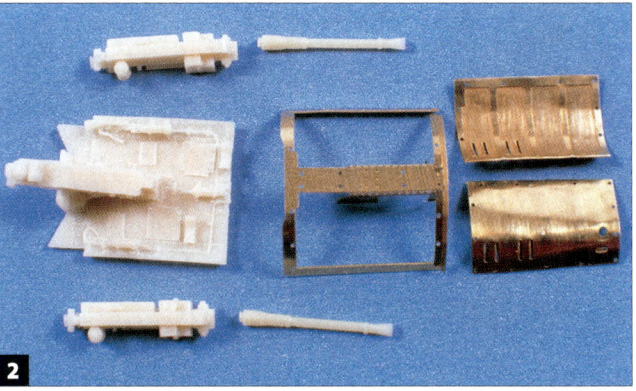
Alfonso modified the nose halves of the Tony to fit the Moskit metal exhaust pipes.

Code-named "Tony," Kawasaki's Ki-61 Hien (Swallow) was the only Japanese World War II production fighter with an in-line, liquid-cooled engine. This resulted in a graceful aircraft, and I just had to have a model of it in my collection. Hasegawa's 1/48 scale kit was easy to build. The pieces fit together well, and it's not an expensive kit. With such a good start, I decided to add a detailed cockpit and machine guns, and improve the engine exhausts with aftermarket parts. I chose 149th Sentai (group) markings from AeroMaster's decal sheet.

Aires interior. Aires' detail set includes resin and photoetched parts that fit perfectly into the Hasegawa kit, **1**. It includes cockpit and back-deck parts, and provides detail for the nose-gun bay as well, **2**. To show them off, I had to cut away part of the top deck and right side of the nose behind the engine.

I painted the interior parts with Vallejo acrylics, **3**. The base color for the cockpit and landing gear wells is a 50-50 mixture of 984 matte brown and 981 orange-brown.

While the fuselage halves were open, I cut out the kit engine exhausts and substituted Moskit aftermarket hollow metal exhausts, **4**. After they were installed, I closed the fuselage, **5**.

The only extra detail I added to the landing gear was copper wire for the brake lines, **6**. I had to fix the trailing edge of the wing where I had gotten a bit too zealous sanding the seams. I used thin strips of plastic and super glue to restore the sharp trailing edge, **7**.

I decided not to try cutting the one-piece canopy apart to pose it open, so I carefully attached it with very tiny drops of super glue, polished it with compound, and then masked it with Tamiya tape.

Airframe preparation. The color scheme I chose was a field-applied camouflage over natural metal. As with any natural-metal finish, the surface of the model must be flawless. I sanded the entire fuselage and wing assemblies so there would be no bumps in the finish. First, I used 1,000-grit sandpaper, then I polished the surface with Tamiya rubbing compound and a cotton ball. This process eliminated seams and buffed out the tiny scratches left by the sandpaper.

Many Japanese factories left their planes unpainted with only national insignia applied. Camouflage and unit markings were often applied in the field, and the quality of the paint and its application were usually poor. That resulted in scruffy looking aircraft. My Tony would be no exception.

I used a European-brand acrylic paint for the underlying natural-metal finish. Whichever paint you choose, it should be fine-grained to resemble aluminum and should be able to withstand masking tape and enamel washes. I tinted several panels with aluminum mixed with a little white or a little black. After three coats, I buffed the finish with a cotton cloth, **8**.

For the field-applied camouflage, I thinned Tamiya XF13 green and varied the density of the paint as I went along the surface

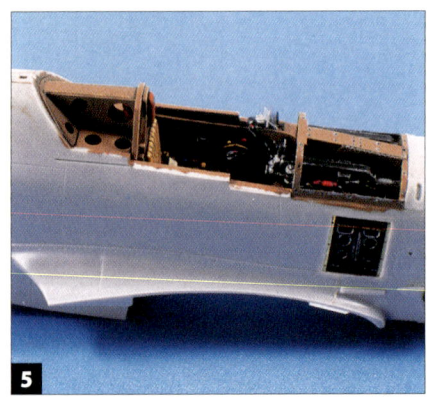

5 The fuselage was assembled and ready for canopy and wings.

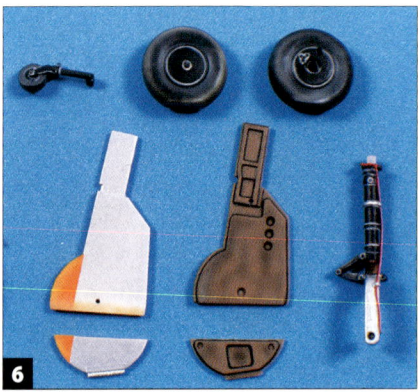

6 Other than a terrific paint job, the landing gear needed only copper wire brake lines.

7 Here's the model ready for paint. Alfonso repaired the trailing edge of the left wing with strip styrene and super glue.

8 The first coats of paint were shiny aluminum, the look of the real aircraft before camouflage was applied.

9 The gray-painted ailerons and elevators, the yellow leading edges, and the red hinomarus were masked and painted.

10 The exhaust pipes were painted, and stains were added along the fuselage.

to simulate the original haphazard paint job. Next, I applied a wash of Tamiya XF55 light gray on the wing spars, then started to "wear" the green off of the airframe. I lightly dabbed a little alcohol with a fine-tipped brush on the areas around the cockpit and wing root. The alcohol dissolved the green paint and revealed the natural metal beneath without affecting it. I continued lightly removing the green paint until I achieved the look I wanted.

Painting the details. I masked and painted the hinomarus (national insignias), fuselage identification stripe, and the yellow wing leading edges, **9**. I painted the bottoms of the flaps and elevators XF14 light gray. When all the paint was dry, I airbrushed a light coat of Micro Gloss in preparation for a detail wash. I thinned black enamel and brushed it into the engraved details and recesses.

I painted the exhaust pipes and stains on the fuselage with a mixture of XF64 dark brown and XF1 black, **10**. Next came decals for the unit emblem on the tail and markings on the main gear doors.

The last paint step was an overcoat of clear flat over the painted areas, keeping it off the natural-metal areas as much as I could. At that point, I removed the masks from the canopy, attached the landing gear, and made a pitot tube from fine stainless-steel tubing.

The finished Tony looks decidedly unkempt; just what I wanted!

THE KAWASAKI KI-61 HIEN ("TONY")

In April 1940, Kawasaki received blueprints and several examples of the Daimler-Benz DB 601A 12-cylinder, inverted-V, liquid-cooled engine from Germany to begin licensed reproduction of this engine as the Kawasaki Ha-40. Manufacture of this 1,100-horsepower engine began in November 1941, with the official name "Army Type Two engine."

Meanwhile, Kawasaki developed two fighter designs to use the new Type Two. The Ki-60 heavy fighter did not go into production, but the Ki-61 Hien lightweight fighter did. It is possible that Allied analysts dubbed the aircraft "Tony" because it looked similar to the Italian Macchi Mc. 202.

Prototypes were tested in simulated combat against an imported Messerschmitt Bf 109E-3, a captured Curtiss P-40E, Nakajima Ki-43 (Oscar) and a Ki-44 (Tojo).

The Tony first flew combat missions in April 1943 in New Guinea with the 68th Sentai and 78th Sentai and was superior to Allied fighters. Maintenance problems resulted in minor redesigns, but the fighter held its own until outclassed by the P-51 Mustang in 1944.

– Alfonso Martinez Berlana

Shoki
SHOWPIECE

Simple updating brings new life to Otaki's old 1/48 scale Nakajima Ki-44 BY FRASER GRAY

Tucked away at the back of most model shops are kits that just don't seem to sell. Their faded, dusty boxes have lost their consumer appeal, which is often reflected in low prices. But if you open some of these boxes, you may find some excellent kits inside.

That's the case with the 1/48 scale Nakajima Ki-44 Shoki (Allied code name Tojo) produced in the mid-1970s by the Japanese company Otaki. The kit is accurate and easy to assemble, with good fit, fine detail, recessed panel lines, clear construction diagrams, and an attractive color profile. The model was later released by Arii, and both kits are still generally

available. I've chosen the Otaki version to illustrate simple techniques for turning any older kit into a rewarding modeling project, indistinguishable from the latest releases when finished.

My only problem was finding detailed references for the original aircraft. I also have an unbuilt Hasegawa Ki-44, and its box illustration shows two holes just behind the pilot's headrest that are missing from both kits. Unable to find any information, I wrote Peter Ferris, an author and historian of Japanese aviation, to find out what the ports were. He explained that the two holes are windows called akarimado in Japanese – literally, "windows to let in sunlight." I represented

Fraser used simple construction and correction techniques to upgrade Otaki's 1970s-vintage 1/48 scale Ki-44 into an eye-catching model that can keep up with the latest and greatest kits. He used the kit decals and painting instructions to complete the model as a Shoki flown by Capt. Wakamatsu of No. 85 Flying Corps 2nd Squadron in 1944.

them with paint on my model, but for others building this kit, I recommend drilling them out and filling the openings with Microscale Micro Kristal Klear.

But it's time to get out the tools and cut some plastic!

Old kits are not always stored in ideal conditions and may have suffered from temperature stress causing the plastic to become brittle. To avoid damaging small parts, I removed them from the sprue with a pyrogravure (a hot knife or wood-burning tool), melting through the connecting stub, and then cleaned them up with a hobby knife.

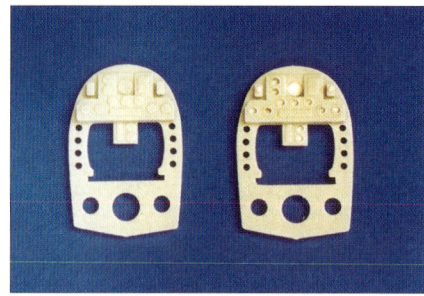

In the cockpit, the kit's main instrument panel (left) was OK but benefited from some improvement. I drilled out the dials and backed them with sheet styrene. After painting, the dial faces can be added either from decals or by painting the dials flat black and removing the paint with a needle to reveal the white plastic underneath.

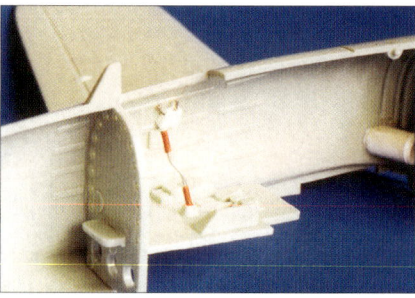

I removed most of the molded detail on the cockpit walls before installing the rear bulkhead and floor. After consulting my references, I replaced the kit's over-simplified control column with a new one fashioned from electrical wire. I removed most of the insulation but left enough to represent the pilot's hand grip.

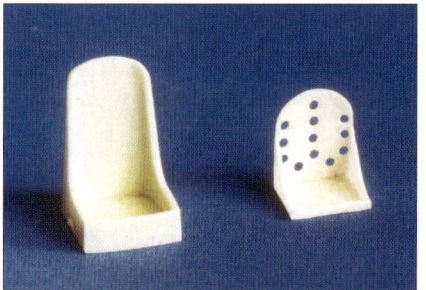

The kit's molded seat wouldn't do (left), so I drew the correct seat pattern on thin sheet styrene and drilled the holes in the seat back before cutting out the parts. The back, sides, and base were glued together with super glue.

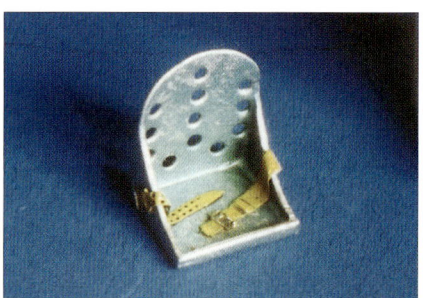

The only aftermarket additions to this project are the photoetched seat belts from Eduard.

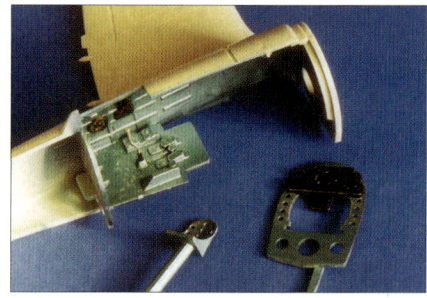

The cockpit interior must be painted before the fuselage halves are joined. I temporarily glued the control panel and seat to cocktail sticks to hold them during painting. On some Japanese aircraft, the interior was painted with a translucent phenolic lacquer called aotake. I airbrushed mine with AeroMaster 1084 aotake, then lightly dry-brushed with Humbrol silver enamel to represent wear.

I opened up the spent cartridge ejection chutes on the wings by drilling out the middle of each port and removing the rest of the plastic with a sharp hobby knife. I used the same process to open up the prominent landing light, then cemented the wing halves together and let them dry overnight.

The next day, the wing seams were cleaned up by wet-sanding with 1500-grit wet-or-dry auto-repair sandpaper. A new front landing light was cut from clear sprue, carved, filled, and sanded to conform to the contour of the wing's leading edge.

After repeated test-fitting and sanding, the light fit. Polishing it with toothpaste (the abrasive kind, not a gel) and automotive rubbing compound restored its clarity. For a final touch of realism, I drilled a tiny hole in the back to simulate a bulb.

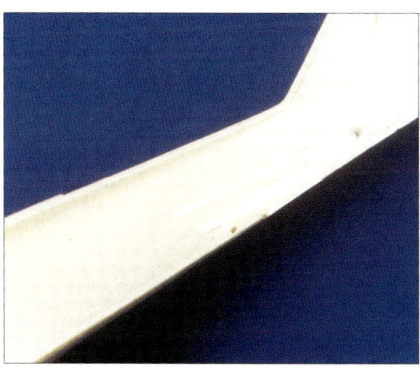

The tail wheel compartment floor is molded into the fuselage halves and is too shallow. I cut it out with my hobby knife and made a new one from .030" sheet styrene. I drilled two holes to locate the tail-wheel leg and retraction rod. Because the new bay is deeper than the original, the leg and rod had to be extended with lengths of styrene rod. The seam between the kit parts and the extensions was drilled and reinforced with fine wire.

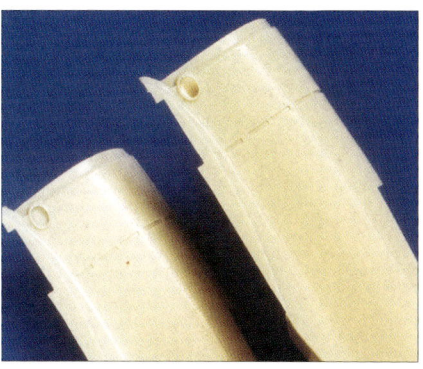

Up front, a prominent feature on the radial-engined Shoki is the exhaust outlets. The kit outlets were too shallow, so I drilled a pilot hole in each one and carefully opened them up with a hobby knife. A length of styrene tube was cemented inside the opening and blended in with Miliput putty. The fuselage could now be cemented to the wings.

I opened up the carburetor intakes on the cowling and installed new intake ducts made from .015" sheet styrene.

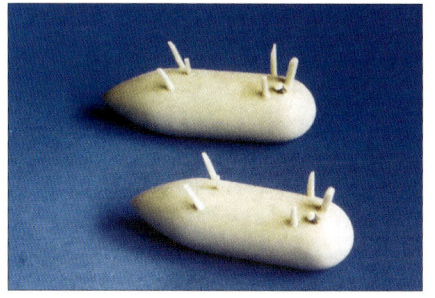

The underwing auxiliary fuel tanks were detailed with new styrene-strip support struts (say that three times fast!), a filler pipe cut from styrene rod, and a styrene-sheet filler cap punched out with a Historex punch-and-die set.

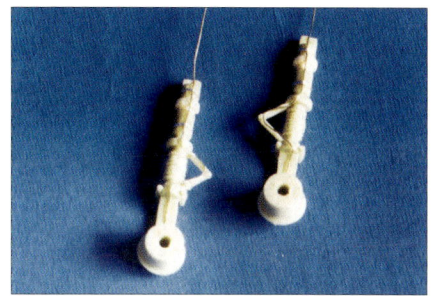

The kit's well-detailed main-gear legs needed only brake cables made from fine wire from a fuse and cable securing straps made from metallic foil from a pack of hobby-knife blades.

Holding the main-gear tires against a hot clothes iron with a pair of tweezers softens the plastic to produce the realistic bulge of a tire supporting the weight of the aircraft. Make sure you clean the iron after use to avoid domestic problems!

The detail on the kit engine is good, so I brushed it flat black, then airbrushed a top coat of silver and gloss black to give it a pleasing metallic finish. I couldn't find any data on the interior color of the engine cowling and firewall, so I airbrushed them dark gray.

After installing the cowling and auxiliary fuel tanks, I corrected the tail-wheel doors by thinning them with an oval file before cementing them in place. The model was cleaned with a cloth moistened with a light solution of dish detergent to remove dust before painting.

I cut lengths of Tamiya masking tape to cover the front of the engine and the cockpit. Humbrol masking fluid was applied over the tape to eliminate any chance of overspray getting in.

I applied an undercoat of gray auto primer, then began creating the worn appearance common to WWII Japanese fighters by airbrushing the entire model with Humbrol silver enamel. I let the paint cure for several days. I airbrushed the wing leading-edge recognition panels and the anti-glare panel.

There is some dispute over the color of the antidazzle panel; some sources say it was black, others mention a dark blue (probably the result of sun bleaching). I compromised and made mine blue-black.

When the paint dried, I masked the panel with more masking fluid and Tamiya masking tape cut into thin strips. I took my time and was as neat as possible; it saved time on corrective painting later on.

The main- and tail-wheel wells were painted the same color as the cockpit interior, then masked before the entire fuselage was airbrushed in Japanese Army Air Force light gray. I mixed twice as much paint as I thought I'd need to be sure I wouldn't run out before finishing the project.

I masked the bulk of the underside with Tamiya masking tape to prevent overspray. The model was ready for application of the upper-surface camouflage pattern.

To get the right effect, there should be a slight gap between the card masks and the surface of the model. I held them in place on the Shoki with Blu Tak. As I applied the JAAF green, I angled the airbrush slightly below the mask and sprayed upward. As the spray struck the mask, it produced an even but soft demarcation line.

I airbrushed the random brown mottling on the upper surfaces freehand. When I sprayed near the mask, I repeated the procedure of spraying upward from slightly below the mask.

To simulate the heavily worn appearance of camouflaged Ki-44s shown in war photos, I used a sharpened toothpick to scrape away the top colors and reveal the silver undercoat. The areas along the pilot's route to the cockpit, around access panels to the engine and armament, and along the wing's leading edges received extra attention.

The same weathering technique worked well on the propeller. I airbrushed the components with Humbrol silver. The blade tips were painted yellow, then masked, and the blades were given a coat of dark brown, with red for the hub. After the paint dried, I picked at the top coat with a blunt hobby knife blade.

The canopy was prepared for painting by polishing it with auto polish and washing it in soapy water. I masked the canopy frames with thin strips of tape and masking fluid and then sprayed the frames with the same color as the cockpit interior, followed by the green-and-brown camouflage pattern to match the rest of the airframe.

I used a photocopier to enlarge the decal-positioning drawings from the instructions to the size of the model, so I could position the hinomarus (national insignia) accurately. The entire model then received several light coats of Future floor polish to produce a glossy surface for the decals. Despite their age, the kit's decals were still usable.

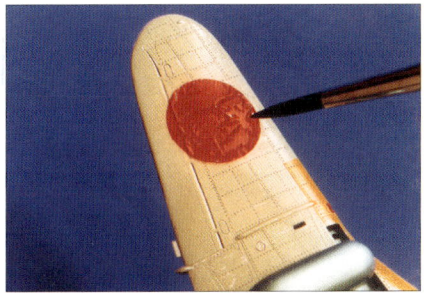

A light application of Micro Sol helped the decals conform to the model. The initial result was alarming as the decals wrinkled up like prunes. As they dried, they wrapped to the surface detail. Where air got trapped beneath the decal, I pricked it with a pin and applied a little more Micro Sol with a fine brush.

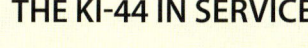
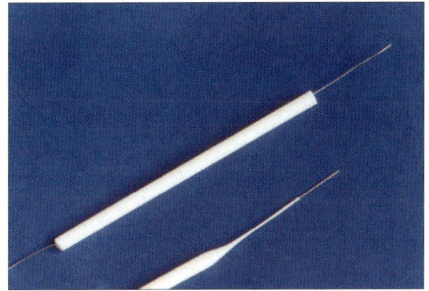

For the pitot tube, I put fine fuse wire inside a length of styrene tube and heated it over a candle. I stretched the plastic tightly around the wire, trimmed off the excess, and glued it to the wing.

THE KI-44 IN SERVICE

After five years of development to create a first-rate air defense fighter, the Ki-44 entered Japanese Army Air Force service in 1942. It was named "Shoki" after a demon slayer from Japanese mythology, whose spirit protected rulers of Japan against invading demons.

Despite a high top speed and good rate of climb, the Shoki was not popular with its pilots because it lacked the maneuverability of other Japanese fighters. It was a match for most Allied fighters, but it proved less effective when sent against high-flying B-29 bombers later in the war. Nevertheless, it served to the end of the war.

– Fraser Gray

Another coat of Future sealed the decals before I airbrushed on a coat of Polly Scale clear flat. Brown and black artist's pastels applied with a fine brush completed the weathering. I finished by installing an antenna array of fine fishing line.

Want to learn more about scale modeling?
Try these!

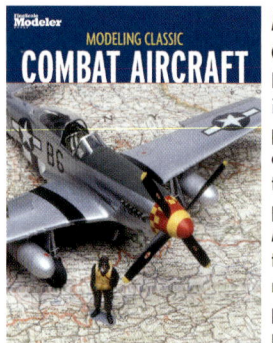

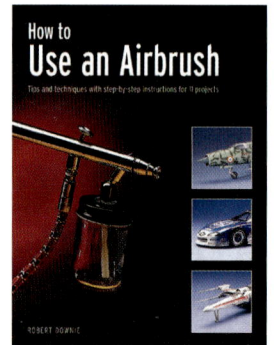
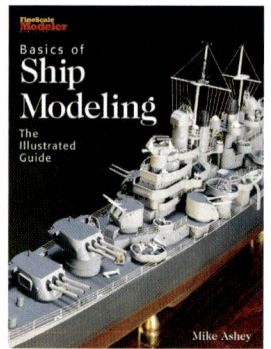